The Enigma of Energy

Nursing Dept...

425 East 25th St.
New York, NY
10010

The Enigma of Energy

Where Science and Religion Converge

VIDETTE TODARO-FRANCESCHI

A Crossroad Book
The Crossroad Publishing Company
New York

The Crossroad Publishing Company
370 Lexington Avenue, New York, NY 10017

Printed in the United States of America

Library of Congress Cataloging-in-Publication Data

Todaro-Franceschi, Vidette.
The enigma of energy : where science and religion converge / by Vidette Todaro-Franceschi
p. cm.
Includes bibliographical references and index.
ISBN 0-8245-1785-7 (pb)
1. Force and energy – Miscellanea. 2. Vital force. I. Title.
BD701.T56 1999
118–dc21 99-24237

1 2 3 4 5 6 7 8 9 10 04 03 02 01 00 99

Dedicated to
Michael

Twenty-seven years,
sharing the laughter
as well as the tears.

Without you,
I could never have done
any of it.

Contents

Acknowledgments

This book is the outcome of a philosophical inquiry that began while I was working as a cardiac critical care nurse and attending graduate school at Hunter College of the City University of New York in 1987. The inquiry formally ended (only formally has it ended) in the culmination of a doctorate degree at New York University in January 1998. So there are a great many people that I could name who have influenced me in one way or another, whether or not they are aware of it. Since a rather long list would be needed, and I have already thanked many individuals in previous acknowledgments, please know that I am thanking all of you once again. I also wish to thank all of the people who I cannot name who have, throughout my life, in some way, shape, or form, contributed to the development of my ideas. I know there have been many.

In the revision work from thesis to book, a few individuals helped with the transformation and need honorable mention. I am convinced that editors are a special brand of people. They can tear you down while building you up and you never feel offended. Probably because they're usually right. The conversion is not a simple one, and former publisher of Crossroad Publishing Company Michael Leach made the transition much easier with his witty comments and honest critique. Rev. Bob Pagliari, who has taken over Mike's responsibilities at Crossroad, tightened things up for me and added more clarity here and there. Thanks go as well to the rest of the Crossroad staff involved in this project and especially John Eagleson, who performed all of the editing and interior design.

For years I didn't go to church, but as time went on, my work drew me closer to religion, and I felt compelled to return to my old parish. When this work was accepted for publication I asked Msgr. James McMahon, former pastor at the Church of St. Brendan in Brooklyn, to review it. He shared it with another priest, Father Joseph Niet Long, who is from Nigeria and just coincidentally happened to be working at St. Brendan's for the summer. But as you will find out a bit later, I don't really believe in "chance" coincidences. Father Joseph was able to fill in some of my knowledge gaps related to the African religious tradition. I owe a debt of gratitude to them both for taking the time to read my work and offer insightful comments.

Elizabeth Barrett, R.N.; Ph.D., professor at Hunter College, School of Nursing, friend and mentor, has through the years inspired me with her ability to knowingly participate in so many things and do them all so well. She critiqued my work on health and healing related to nursing. Marty Dornbaum, director of the Media Center at Hunter College Brookdale Campus, was especially enthusiastic and critiqued a whole draft of the book. Both of them offered valuable feedback.

This book would never have taken on the many dimensions it has without the dedicated assistance I had from various individuals who graciously critiqued my doctoral work. Peter Simpson, Ph.D., of the Philosophy Department at City University of New York Graduate Center, was my consultant on Aristotelian philosophy while I was a student. He spent an inordinate amount of time reading my work, offering insightful comments, and answering my many "just one quick question" e-mail queries. I still bother him on occasion. Andre Adler, Ph.D., of the Physics Department at New York University was kind enough to read and comment on some of my doctoral work related to physics. As I swam in water over my head from time to time with those such as Aristotle and Einstein, it was reassuring to have experts in these other fields provide some feedback.

In particular, I owe a special debt of gratitude to Thomas B. Colwell Jr., Ph.D., Joanne K. Griffin, R.N.; Ph.D., and John R. Phillips, R.N.; Ph.D., all of whom were on my dissertation committee at New York University. Tom, whose expertise lies in the philosophy of ecology and environmental education, told me early on that everyone philosophizes or at least should, so it was okay that I do so. It was he who introduced me to Collingwood's *The Idea of Nature*. Joanne was my most ardent editor during the thesis writing. She offered countless invaluable insights, particularly in the area of religion. John, who started off as one of my professors, has become much more than that. He has been, and continues to be, a continual source of inspiration. If not for the many, many hours of intellectual discourse I had with John on most of the topics that follow, much of this work would never have come to fruition. Both John and Joanne also critiqued parts of my final drafts for this book.

Grateful appreciation goes as well to those individuals whose work I have read and synthesized to come up with my picture of the whole. Many of them are no longer *physically* with us — but their essence lives on in us who have come after. Others are still physically here, and some will no doubt disagree with what's on the pages that follow. That's okay. Sometimes the best kind of thinking emerges from controversy. And one of my foremost aspirations for this book is to stimulate thought and reflection, the metaphysical kind of reflection that most of us don't frequently engage in, probably because we are too busy with our day-to-day survival strategies.

None of this would have been possible without my family. Love, gratitude, and admiration go to my parents, Charlotte and Ernest Todaro. *All* of my good qualities are related to my genetic inheritance from both of them and/or their devoted parenting. My mother is a registered nurse, and the impression her life and career have made on me is obvious, for I became a nurse as well. From her, I either inherited, or learned, compassion — a mandatory quality for any professional service. She has always been there for all of us whenever we needed her. I must have inherited my father's inclination for philosophy. We spent many hours talking philosophically about "energy" and other things. He was the real impetus behind my going on with my education.

I had originally thought to dedicate this work to my father, as he really has been the wind beneath all of our wings in my family, but I know that he would point out the obvious. Like most daughters, I've never done anything of import without asking the opinion (and/or permission) of my father. He has a different celestial address now, but I can still hear his words and thoughts ever so clearly. I guess that's one of the ways you never really lose those you love when they die. You always think, "What would he or she say," before you do anything, and you usually base what you do on what they would have said.

In fact, a very meaningful coincidence occurred while I was lamenting over the question of whom to dedicate this book to, and it provided an answer to my dilemma. Actually, you could say my father provided an answer. I'm not ready to share this one, yet, even though I do write about meaningful coincidences in this book. Perhaps one day I'll share this one as well. Although I could have interpreted the meaning of this incident in many ways, when it happened I just *knew* unequivocally what I should do. I heard my father telling me, clear as day, as he had on numerous occasions over the years, that I had my husband to thank for many things. And he would think it wrong of me had I dedicated this book to anyone other than Michael.

No amount of love and gratitude could repay Michael for his selfless support and encouragement throughout the years I have labored over this work. He cleaned house and kids, did homework and laundry, and ate take-out for dinner, on an all too regular basis. He frequently tripped over my books and files dispersed throughout the house and tolerated it all with a smile (well, usually). He bought me one computer after another every time I needed an upgrade. He spent far too many eves and weekends alone while I stayed in my den reading and writing. He came up with the word "enigma" for the title on one of the many, many nights that I kept him awake till dawn talking about "energy." I am very fortunate to have him as my partner in life.

Boundless love and gratitude also goes to my three daughters, Virginia, Laura, and Tara for understanding how important this has been to me. I deprived them of much quality family time while I locked myself in my den to work. Grateful appreciation goes as well to my siblings, Celeste and David, and grandchildren, little Michael Ernest and Vidette Charlotte. All of them contributed to this work in their own unique ways.

Introduction

This book is the culmination of over a decade of research concerning the idea of energy. I had set for myself the task to get beyond the everyday use of a word, perhaps one of the most widely used terms in existence, in order to elicit some universal meaning for a phenomenon that is the basis for our being and becoming. The search for universal meaning often compels the seeker either to give up the quest or to focus more narrowly on a given subject. I could do neither. Driven by an intuitive feeling that what has been imbedded in the idea of energy is something immensely important for everyone, I have tried to look at the whole picture. I am a nurse by profession, and in nursing we try to understand the whole. The whole encompasses not only the human being but also the environment, all inanimate and other animate things. One without the other has no meaning. Accordingly, the purpose of this book is to shed light on what nurse scholars often refer to as the pattern of the whole.[1]

In order to examine the whole you can never set limits. So, what started out within the narrow realm of the health-related sciences quickly broadened into philosophy, physics, psychology, and eventually theology. Consequently, I can perhaps use educator Gabriel Moran's words to defend the manner in which I went about this inquiry. In his book *A Grammar of Responsibility,* he notes:

> A writer has to be either audacious or naïve to write a book that crosses as many academic divisions as this one does. I am fully aware that scholars rightfully protect the boundaries of their disciplines from presumptuous intruders. My only defense is that, having decided to track the term . . . I followed where it led.[2]

While I have also gone well beyond my own discipline, throughout my wanderings I have been most humbly and sometimes quite painfully aware that I am not an expert in these other fields.

My interest in the concept of energy began with an innocent curiosity about the idea of energy conservation in critically ill adults. Working in a cardiac intensive care unit, I became increasingly perplexed with certain procedures nurses and physicians performed on a routine basis. We woke patients up during the night to obtain hourly vital signs. We measured all body temperatures rectally because it was believed to be the most accurate measurement even though it often created increased emotional and physical stress, particularly for the male patients. Daily weights were the best indicator of fluid balance. So we weighed patients daily even though we usually had to use a bed scale, which meant putting the patient through all kinds of positioning maneuvers in order to get a weight. These procedures seemed to result in increased and, what seemed to me, unnecessary energy expenditure for the patients.

The clinical concerns that I had years ago have now been resolved to some degree. Vital signs are for the most part monitored by various devices that do not require waking patients during the night. Rectal temperatures are no longer routinely taken in most clinical areas, and daily weights have become much easier to obtain with advanced technology since most facilities now have bed scales built into the beds. But over time my interest in the philosophical implications of energy has continued and grown into a passion.

All along I had this nagging feeling that something in the very essence of the idea of energy needed to be examined. Somewhere along the way I intuited that in health care we couldn't possibly be expending or even conserving human energy. Rather, I came to think that what we do is assist individuals to transform or change their unique energy form and in so doing, purposively actualize inherent potentials.

In the early part of the twentieth century French philosopher Henri Bergson gave a lecture on philosophical intuition that expresses how I felt while performing this research.[3] He likened intuition to a shadow that follows the philosopher while he philosophizes. The shadow directs the individual to discount some ideas while retaining and further exploring others. It is often this kind of intuitive feeling which guides a person in the search for answers to metaphysical questions, such as the one posed here. I would call it knowing something while not knowing what it is.

Bergson said that each person who philosophizes necessarily does so within the context of a larger whole:

> Doubtless the problems which the philosopher worked upon were the problems which presented themselves in his day; the science he used or criticized was the science of his time; in the theories he expounds one might even find, by looking for them, the ideas of his contemporaries and his predecessors. How could it be otherwise? In order to have the new understood, it must be expressed in terms of the old.[4]

Not surprisingly, my interest in the phenomenon of energy and its conceptualization over the course of history grew further when I read the work of the ancient Greeks, most notably Aristotle. It was apparent to me that historical as well as contemporary work was crucial to an understanding of the idea of energy.

My goal was to illuminate and perhaps answer the question "What is energy?" by comparing and contrasting scientific discoveries along with the philosophical implications that have occurred over time. Yet, as Bergson so artfully noted, what may at first glance be seen as a clarification of ideas within a given context may on closer examination reveal the articulation of a new idea:

> So many partial resemblances strike us . . . that we are tempted to recompose the philosopher's thought with fragments of ideas gathered here and there. . . . But the illusion does not last long, for we soon perceive that in the very places where the philosopher seems to be repeating things already said, he is thinking them in his own way.[5]

I admit, I *am* thinking and writing things in my own way, but it does seem to me that the ideas that will be brought to light on the following pages have been there all along. They can be found most readily with an open mind in the words of the great mystics, scientists, and philosophers throughout the ages. So what I have done is synthesize and extract from various sources the essence of an idea.

Every culture, every philosophic tradition, has been talking the same talk, albeit in its own unique way. What's unfortunate is that we haven't been able to recognize it, we haven't been able to grasp the larger picture — the whole. So while we've been unknowingly participating in this universal communal process, talking the same talk, for the most part we haven't been *living* the talk.

At the onset of this project I knew that my journey would take me well beyond my particular discipline, but I never expected that it would to lead me into religion. Frankly, never having been devout, I didn't realize that philosophy and religion were, in many cultures, inseparable. The universal idea of religion, or perhaps I should say spirituality, smacked me in the face while I was writing about the origin of the universe and life in chapter 3. At that time it all came together for me. The essence of the enigma of energy is of a spiritual nature. Energy — the phenomenon — is not bound *in* the sciences, and although the *concept* of energy is obviously important to the sciences, it has perhaps even more meaning in the context of religion.

The more I worked on this project the more I became aware that somehow science and religion were converging. It was never my goal to merge these two seemingly disparate areas; in fact, when my search led me into religious realms of thought, I tried hard at first to stay clear of them. But it was impossible to do so. Anytime I came across literature that was related to an idea of energy there were implicit or explicit spiritual overtones. Most surprising was the abundance of spiritual ideas found in physics. It seems that you simply cannot talk of wholeness or oneness without getting into some kind of religion.

Inevitably my examination of the concept of energy led me to two disparate ideas: energy can be viewed as part of a causal process where energy is used, gained or lost, or it can be viewed as a phenomenon, a thing which is always changing. The word "phenomenon" as I use it refers to an actual extraordinary thing, but more importantly, *one* thing, which is constantly transforming and manifesting itself in different ways. So that although we perceive that it exists, there are aspects of the thing that are ultimately unknowable. These two different ideas of energy have pervaded the course of history, leading in turn to different worldviews which I hope to give you a glimmer of as we go along.

Energy is an enigma. Yet, all of existence as we know it is contingent upon this mysterious thing called energy. And every day, as technology advances and the sciences progress, it becomes even more paradoxical. However, by delving into the background of the concept, not merely from a physical science viewpoint but also from a philosophical viewpoint, we may begin to see a coherence that previously was not evident.

Journalist John Horgan, in a lively discussion about postmodernism, or "ironic science," notes that "the job of the literary critic is thus not to pin down the true meaning of a text — an impossible task, but to invent new meanings, ones that challenge and provoke further dialogue."[6] This work may partially fall into the category of literary criticism since given the subject matter at hand it may be impossible to "pin down the true meaning." Nevertheless, it should be possible to stimulate further thought and dialogue pertaining to the enigma of energy.

On the following pages I have tried to bridge the gap that has developed between the sciences regarding the conceptualization of energy. In doing so, I did not limit this inquiry to any particular focus or scientific discipline. And when we get right down to it, we find that ideas of energy are pretty much the same in every science and culture, and in a way *they remove the boundaries* thought to exist.

What I hope to give you is an expanded awareness which I myself experienced during this inquiry and which has continued to pervade my daily living. I no longer view energy as part of a causal process which enables (or disables) me to do things. Rather, I see energy as a thing, a phenomenon, which is always purposively changing. We are but manifestations of a whole, a whole that by its very nature is both knowable and at the same time unknowable. All are somehow one, and I think on some level we have always known it. To that end, I invite you to open your mind and heart so that I may share with you what I have learned while researching this wondrous mystery.

The Enigma of Energy offers authentication for many of the conscious, and sometimes subconscious, ideas that humans have had since time immemorial — notions such as: all is essentially one and the purposive actualizing of potentials is an inherent characteristic of every living and nonliving thing. The ideas brought to light herein, of necessity, ultimately must lead to the conviction that there is *a reason for it all.* There is a *definite* pattern in the whole.

Chapter 1

The Enigma

> Herein is especially apprehended the unity of Nature, — the unity in variety, — which meets us everywhere. All the endless variety of things make an identical impression. Xenophanes complained in his old age, that, look where he would, all things hastened back to Unity. . . . The fable of Proteus has a cordial truth. A leaf, a drop, a crystal, a moment of time is related to the whole, and partakes of the perfection of the whole. Each particle is a microcosm, and faithfully renders the likeness of the world.
>
> — Ralph Waldo Emerson[1]

Ever since humans developed into sentient beings we have pondered why we are here and what, after all, life is. Metaphysical questions such as, Is there purpose in the universe? or, Is everything we see around us just a chance occurrence? have occupied men and women throughout the ages. Our existence is an enigma, and, on a far greater scale, so too is everything in and beyond this universe. The concept of energy, nay, the phenomenon of energy, and its consequent conceptualization in the various sciences, is perceived herein to be the source of this enigma.

The concept of energy has been dominant in the history of ideas, and, more importantly, it has played a significant role in the paradigmatic shifts that have occurred over time in the various sciences. However, the word, and hence the concept, is a vague nuance rather than a formalized idea of what this thing — energy — is. And energy is a "thing" in the modern sense of the word. Energy is a phenomenon, an extraordinary thing, and as such it has been conceptualized in various sciences and in several ways.

Contrary to the popular belief that energy can be or is simply defined, on the following pages it will become clear that our modern, commonly accepted definitions of energy are insufficient to describe this phenomenon. A few years ago an article appeared in the *New York Times* on the enigmatic nature of complexity. In it the science writer George Johnson notes:

> Dr. Stein compares the challenge of defining complexity to that confronting early 19th-century scientists as they tried to get a grip on a mysterious concept called energy. Today, people take energy so much for granted that it is hard to appreciate how abstract the concept really is. "Many people had a pretty good idea what energy did and how it behaved," Dr. Stein said. But energy was not really understood, he said, until people came up with a precise definition. The result was the laws of thermodynamics.[2]

While energy is an abstract concept that is commonly taken for granted, it — as a phenomenon — doesn't seem to be any less mysterious now that we have the laws of thermodynamics. If anything, it might be even more enigmatic.

The first law of thermodynamics is also known as the conservation of energy principle. In his book *Understanding Thermodynamics,* H. C. Van Ness notes:

> What is energy? One might expect . . . a nice clear, concise definition. Pick up a chemistry text, a physics text, or a thermodynamics text, and look in the index for "Energy, definition of," and you find no such entry. . . . Every time they have an opportunity to define energy, they fail to do so. . . . Whatever it *is,* one thing we know about energy is that it is conserved. That's just another way of saying that we believe in the First Law of Thermodynamics. Why do we believe in it? Certainly no one has proved it. On the other hand, no one has been able to find anything wrong with it. All we know is that it has always worked in every instance where it has been applied, and we are happy with it simply because it works. Why does it work? We haven't the faintest idea; it's just a miracle of nature. The conservation law is a *description* of how nature works, not an *explanation.* Fortunately that's all we need.[3]

Fortunately for me as well, since I would be hard pressed to go into any technical details on this subject. But having read this I realized that the nature of the enigma of energy is enhanced by the various, and often inconsistent, meanings attributed to the word "energy."

Energy: The Word

Think of how often we hear and use the term "energy." Most people use the word so frequently we are not even aware of it. Yet, ask for a definition and many individuals draw a blank. For most of us, the word "energy" requires some thought or reflection before we can articulate a meaning, that is, unless one happens to be a physicist, biologist, or chemist. Then at least we should have an idea of how energy is defined in our particular discipline. The great philosopher Ludwig Wittgenstein's thesis that there is no one thing a word stands for is substantiated by the term "energy," for over time the conceptual development of energy has branched into numerous directions with resultant multiple meanings. Each science uses the word in a different context and sometimes in more than one way. It is likely that this has contributed to confusing and conflicting use of the term.

When energy is discussed in the context of a particular discipline, the word usually signifies a particular process or phenomenon or, at the very least, a form of something specific. For instance, biochemists or physiologists often define energy in terms of metabolic requirements, and they might speak of the energy we use internally, adenosine triphosphate (ATP). Scientists in various disciplines speak of energy as the ability to do work and often distinguish different kinds, like potential and kinetic energy. In addition, those who are well versed in what we call the "hard core sciences," for example, physics, define energy with mathematical equations that make little or no sense to the ordinary individual. I am one of those ordinary individuals; mathematical constructs make no sense whatsoever to me.

In the health sciences the word "energy" is used in many ways, each specifically related to the human body. Traditionally, physicians discuss energy in terms of expenditure of a particular body part or function, for example, myocardial energy expenditure is said to be signified by the heart rate. When the heart rate increases, myocardial energy expenditure is said to increase as well, essentially meaning the heart pump is working harder and requires more oxygen. However, there are also physicians who speak of energy as a spiritual phenomenon. Nutritionists focus on caloric energy requirements, and physical therapists focus on the body's muscular ability to use energy resources in order to perform physical activity.

Nurses may use the word "energy" in various contexts not solely describing specific phenomena. Some nurses speak of the energy requirements of the person while others use the term in a more abstract fashion. A case in point is Martha Rogers, a nurse theorist who defined human beings and their environments as distinct energy fields.[4] Psychologists often refer to the energy of the psyche or mind.

The Underlying Ideas of the Phenomenon

Take a moment to reflect upon your own use of the word "energy." How often do you say, "I don't have the time or energy to do this or that," or "I feel energized after exercising, or after eating that meal." Reflect upon how often you might think of energy by way of feelings. For example, you might feel that you need more energy, or that something is draining you of energy. On the other hand, a particular activity, person, place, or thing seems to give you more energy. You will find that energy, as an idea and as a phenomenon that you experience, plays a major role in your thinking and feeling on a day-to-day basis.

Yet, there is also deeply imbedded in the idea of energy an allusion to the transformational ability of things; that is, things change. Underlying the idea of change, especially in relation to human beings, is the somewhat still Aristotelian notion of actualizing potentials. So that in health care today, we often speak of helping people to actualize their unique potentials to be happy and healthy. In fact, a number of psychologists such as Kurt Goldstein, Abraham Maslow, and Mihaly Csikszentmihalyi, have developed theories that at a fundamental level address the actualizing of our human potentials.[5] In a similar vein, in nursing, we have a theory of "power as knowing participation in change," developed by nurse Elizabeth Barrett, which addresses the actualizing of our potential for health and healing.[6]

In the various health care disciplines we often speak of people being able to transcend physical impairments, optimizing health, and expanding consciousness, all notions which seem to be inextricably interwoven with an idea of energy. And notably in modern physics, scientists speak of the seemingly inherent ability of elementary particles to actualize or manifest certain potentials while others lie dormant.

Hans Selye, a noteworthy physician, acknowledged:

> In textbooks the definition comes first and leads you to the concept; the reverse is true in actual life. Be it ever so vague, you must first have a concept — derived from observation and symbolized by a name — before you can even try to delimit it more precisely by a definition.[7]

One can conceptualize and define energy as a particular thing, although the definition will vary tremendously, depending upon the context. Yet, underlying this multiplicity of definitions or contexts is the fact that energy is a phenomenon.

Energy can be categorized and described as a thing, and as such it is open to description and explanation. However, often each science focuses on a distinct aspect of energy, frequently leading to the equivocal idea that what we have are many phenomena rather than one universal phenome*non* with multiple facets.* For instance, many individuals frequently use the terms "energy" and "matter" in the same sentence to refer to two separate entities. Conversely, on occasion, we also refer to both matter and energy as if they were one and the same thing.

Historian of science Theodore Feldman noted: "Around 1890 Wilhelm Oswald (1853–1922) advocated that energetics be substituted for the kinetic and atomic theories as the foundation for all physics. Not matter, he claimed, but energy was the sole real substance in Nature."[8] Oswald was certainly on the right track, since Einstein came along later announcing that matter and energy were indeed interconvertible. Nevertheless we often slip back into the old tradition of speaking, thinking, and writing as if energy and matter were distinctly separate and different entities.

*To emphasize energy as *one* thing rather than many things, I frequently italicize the final syllable of the word "phenomenon."

Concepts in General

In one of his many seminal works, Einstein discussed, as he often did, the concept of space; "The Problem of Space, Ether, and the Field in Physics" is particularly interesting as it can also be rightfully considered a lecture on the evolution of a concept.[9] Einstein pointed out that all scientific concepts are derived from what he called prescientific concepts, that is, an original word expressing something that has over time become an idea without a clear relation to experience:

> We have, so to speak, forgotten what features in the world of experience caused us to frame those concepts, and we have great difficulty calling to mind the world of experience without the spectacles of the old-established conceptual interpretation. There is the further difficulty that our language is compelled to work with words which are inseparably connected with those primitive concepts.[10]

In trying to come to a clear distinction of space as a prescientific concept Einstein noted the difficulty of conceptualizing ideas, especially those related to intangible things. Since space is not something tangible and perceptible in sensory experience, it is difficult to define it without looking at space in relation to some solid physical thing.

Einstein noted that the concept of space presupposes knowledge of a solid physical thing. When you see two tangible things, for instance, a chair and a table, you will note an area between the two things that is empty, or at least it seems empty according to your sensory experience. The "empty" interval between the two things then becomes a spatial relation that is just as real as the chair and the table. Thus, as Einstein pointed out, the prescientific concept of space was just as real as the tangible things with which it had a distinct relation. So too, with the phenomenon of energy. It is a thing, but not as tangible as a chair or a table. And, that's because, in a strange way, it *is* the chair, the table and the empty-looking space in between.

A number of years back I spoke with a professor of physics concerning my research. For a little over two hours we tossed around ideas concerning energy. It was immense fun for both of us, yet our ideas of energy differed greatly. He in fact argued that what I was investigating was not "energy." According to him energy is essentially part of a process, not a thing. The example he gave was that of combining oxygen with hydrogen. These two elements alone, he said, are not energy, until they combine to form water. Energy is part of the actual combining process and not the oxygen, hydrogen, or water per se.

I acknowledge that perhaps, if this is so, energy is not an appropriate term for what I have in mind. The energy that I am writing about is referred to and has been conceptualized in a multiplicity of ways, not just as part of a process or the ability to perform work. And it is intricately tied up with metaphysical ideas that have been debated among philosophers, scientists, and theologians ever since humankind acquired the ability to question the meaning of it all.

To begin at the beginning is to begin with Aristotle's conceptualization of *energeia.* But having already performed the research into the concept of energy, it is far easier to begin with what I think of as the end: the findings. Energy is not just part of a process which enables or disables things to change. It is a phenome*non,* meaning not only *a* thing, but *one* very extraordinary thing. One thing that is always transforming, always changing, and thus it never stands still. Ironically, a glimmer of this idea is breathtakingly apparent in the philosophies of the ancients. So that is where we will begin.

Ideas of Energy in Antiquity

In the East, the ancient Chinese held that the universe was a dynamic entity filled with continuous cyclic flow and change. The motion of these cyclic patterns is expressed by the concepts of yin and yang. The yin is the female, dark, quiet and resting, intuitive component of change and is associated with the earth. The yang is the male, light, strong, creative, active component of change and is associated with heaven. Although polar opposites, together the yin and yang comprise life, where there is a continual harmonizing of both cycles of change.[11] These two concepts indicate the underlying Tao (way) or pattern of everything in the universe. The inherently changing nature of all things is held in balance or harmony by a continuous flow of *ch'i*.

Some individuals refer to *ch'i* as "ether" or "air," while others specifically link it with material and vital force or energy.[12] As with all translations one must bear in mind that there are multiple renderings of the meanings of various words. Most often *ch'i* is taken to mean energy, particularly life energy. But the ancient Chinese made no distinction between the cosmos and living things. For them, everything was, in a sense, alive, and all things arose from one common ground. One certainly gets the feeling that the ancient Chinese meant something more than simply a tangible material substance.

Lao Tzu, who is believed to have been an elder in the times of Confucius (fifth and fourth centuries B.C.), spoke in great depth about the "One" from which all else arose.[13] He perceived a distinct and yet indescribable pattern in things which he commonly referred to as "Dao" (Tao; the way). The Confucian canon notes that heaven and all other things are essentially one. Much later on Chang Tsai (tenth century), wrote, "the universe is one; its manifestations are many."[14]

The ancient Indian philosophies also talked about the idea of an essential oneness. Indian philosophers Sarvepalli Radhakrishnan and Charles Moore note that "almost all Indian philosophy believes that reality is *ultimately* one and *ultimately* spiritual."[15] Even in the earliest works there is evidence of a belief in unity. But there was also an awareness of the orderliness in nature, equivalent to the Chinese idea of pattern (Tao). The term *Rta* was used to express an eternal law which accounted for the orderliness underlying all things.

Prana, a term that has been used in the ancient Indian tradition for over five thousand years, denotes a universal or vital life energy. Often translated as "breath" in Indian works, the concept of *prana* is a central one for principles of healing in Indian Ayurvedic medicine, which was developed from 1200 to 800 B.C.[16] Today adherents believe that one can regulate and manipulate the flow of *prana* in order to maintain or restore health using various alternative health care modalities, such as healing touch.

Although the terms used in each philosophic tradition differ, clearly the ancient Eastern philosophers believed that some *one* thing underlies existence. The common belief that there is a vital energy (or force) unique to and inherent in things formed the basis for vitalism, a modern Western philosophy which was held in great esteem by many leading philosophers such as Henri Bergson. The concepts of *ch'i* (Japanese *qi*) and *prana* are still prevalent in modern Eastern philosophy and have significance for many modern-day practitioners of alternative health care modalities.

In the West, the ancient Greeks formed various theories concerning the stuff of which things are made, and many of them believed that there is some one thing underlying all that is.[17] Thales of Miletus in the seventh and early sixth centuries B.C. postulated that all natural things are composed of one universal substance, which he presumed to be water. Later in the sixth century B.C., Anaximander revised this hypothesis by stating that the substance of which everything is made is *aperion,* generally taken to mean some one thing which is infinite or boundless. Anaximenes (sixth century B.C.), professed that air was the source of all things. He also argued that nothing can come from nothing; everything has to come from something. In the latter part of the sixth century B.C., Heraclitus argued that change was the fundamental principle of the universe and that all things were composed of one thing, fire.

Parmenides (500 B.C.) was perhaps the most significant of the pre-Socratic philosophers. Using deductive reasoning, he argued that "to be," or exist, means that all is one. Contrary to Heraclitus, Parmenides argued that being was imperishable (eternal), unchangeable, and yet completely unified in nature. Interestingly, R. Bruce Lindsay, a distinguished historian of science, interpreted Parmenides' idea of change as being supportive of a modern idea of energy as *constancy in the midst of change.*[18] Of Parmenides' philosophy he says: "His emphasis on the idea of unity in nature and his apparent feeling that the common emphasis on change in experience is an illusion can be interpreted as supporting the notion of constancy or conservation."[19]

Empedocles (fifth–fourth centuries B.C.), who is said to have been a student of Parmenides, held that there were four principle elements — earth, air, water, and fire, which composed all things. The four substances are the only real entities; everything else, including all natural creatures, are only chance occurrences resulting from random combinations of these four elements. Empedocles' idea of random combinations of the four "real" elements sounds much like Darwin's theory of natural selection and was strongly argued against by Aristotle later on in antiquity.

Another allusion to an idea of energy is found in the theory of atomism. Commonly accredited to Democritus (460–370 B.C.), atomism held that everything was made up of atoms — minute particles that are constantly in motion. The atomists believed that the movement of atoms was random, having no underlying pattern or direction. In other words, they believed, like Empedocles, that only by chance occurrences in nature did things come into existence. Consequently materialism — the belief that everything is simply matter — was born, and to this day it has continued to have a place in science.

An idea of paramount importance arising from the atomist theory was the notion of empty space. Atoms were believed to be eternally in motion in a void. Unlike Parmenides and Anaximenes, who both held that nothing could come from nothing and thus there could be no such thing as nothingness, most of the atomists accounted for the movement of atoms by assuming that they moved in empty space. This notion of a void was cause for debate in later times and is significantly related to modern ideas of energy.

The philosopher Lucretius (approximately 99–55 B.C.) held that "substance" in its primordial form was eternal. He saw a fixed nature in things which accounted for why things became what they did:

> But, since produced from fixed seeds are all, each birth goes forth upon the shores of light from its own stuff, from its own primal bodies. And all from all cannot become, because in each resides a secret power its own.[20]

Hence, for Lucretius, in each thing there was an inner principle of change for the thing to become what it did. He did not perceive change to be the random movement of atoms, rather, the movement of things was determined by an innate power. Lucretius also alluded to the idea of energy conservation when he said: "naught of what so seems perishes utterly, since Nature ever upbuilds one thing from another, suffering naught to come to birth but through some other's death."[21]

Lesser known ancient cultures, such as the Native American Indians, Africans, and Aborigines also had a strong sense that all is one. Most of their beliefs were never written, but were passed on in oral teachings from one generation to the other, and thankfully many of their ancient beliefs have carried over into present day. Recently I asked someone who is knowledgeable about Native American Indian philosophy to recommend a book for me to read, and she suggested Dhyani Ywahoo's *Voices of Our Ancestors*. My hair stood on end as I read this beautiful work based on the oral teachings of Native American Indians thousands of years ago. It's here. The ancient American Indians were profoundly aware of the interrelation of all things in nature, often expressed by the word "harmony." In modern times they continue to teach the knowledge and beliefs of their ancestors.

The ancient Africans also valued nature and had a strong sense of community, of oneness. Even today in most African tribes this communal sense pervades their daily living. Indeed, it is said that in Africa there is no need for orphanages or nursing homes. Everyone takes care of their own and others when need be. In anthropologist Colin Turnbull's account of the Pygmy life he notes that children call all adults who are of the same age as their parents "mother" or "father," and all others who are older "grandparent."[22] Because African children are taught to value oneness and community, they have a keen respect for all elders. And seeing that today the word "respect" often seems to be missing from our children's vocabulary, I can't help wondering that maybe we are teaching our children the wrong things.

The ancient Australian Aborigines believed that all land and living things are sacred and inseparable. Songs and myths, as in the Native American Indian and African traditions, have carried their beliefs into modern times. But their art forms best reflect the spiritual dimensions of their culture.[23] They describe without any words how they perceived unity and purpose underlying all nature.

Summary

Ancient philosophers alluded to the idea of energy as a phenomenon before the term was ever coined. Many of them touched upon the idea of the transformational, changing nature of things, and that perhaps the source of all things is eternal and, in a sense, all is one. It is only now at the completion of my doctoral work that I find myself compelled to become more familiar with the ancient and modern philosophies of all people, having become acutely aware that the ideas arising from this inquiry are ubiquitous.

Across all cultures and seeming boundaries there exists a universal notion that will grow ever more apparent as we examine the enigma of energy. There is an idea of oneness that can and in fact does transcend the seeming differences of all peoples. I am convinced this idea is universal, and yet it seems we do not have all the particulars, although perhaps we have knowledge of what Aristotle would have referred to as the First principles — truths that are for all intents and purposes indemonstrable. We just intuit that they are essential truths, somehow or another.

Rather than try to capture the development of this concept in its entirety, which would have been an impossible task, in the next chapter I have focused on selected aspects of the historical evolution that have made significant contributions to the idea(s) of energy. The focus is predominantly on the Western viewpoint, partly due to the fact that I am more familiar, and hence comfortable, with Western ideas, but also because of the voluminous nature of the Eastern philosophies. In addition, I have refrained from going into the technical aspects of this history, since the point emphasized here is an idea which, I believe, goes far beyond mathematical formulas.

Chapter 2

A *Very* Brief History of Energy

> Often it is not so much the existence of things that we *do not know,* or about which we are too uncertain, that handicaps our research, but the existence of things we *do know* and about whose interpretation we are quite certain — although they may turn out to be false. Lack of equipment, or even lack of knowledge, is much less of a handicap in original research than an overabundance of useless materials or useless (and sometimes false) information cluttering up our laboratories, our files, our desks, and our brains. — HANS SELYE[1]

To ask the question, "What is energy?" is like asking someone to tell you what the sky is. Every person will give a different answer; some will be concrete, and some esoteric. So we must ask how the concept of energy is defined and then try to see how and if it relates to our everyday experience. But scientific books do not provide an answer to the question, and dictionaries are of limited value. Have you ever walked through a garden maze or one of those mirrored houses in an amusement park? Trying to capture the essence of the concept of energy is pretty much like being in a maze and having a great deal of difficulty finding the exit.

Cosmologist Carl Sagan wrote in his novel *Contact:*

> Even if the Message cycles back and even if we completely decrypt it, how good could the translation be? You know the opinion of Cervantes? He said that reading a translation is like examining the back of a piece of tapestry. Maybe it's not possible to translate the Message perfectly.[2]

My maze-like search was complicated by multiple translations of historical texts in which the concept of energy developed. And after all this, I can honestly say that Cervantes was right. You can never translate a message perfectly, as the abundance of different interpretations will indeed attest, and so you can never really know for sure what the original thinkers were thinking. Yet, as noted so poignantly by Morris West in *The World Is Made of Glass:* "There is a pattern in all this which I begin slowly to understand and which I am trying to codify. The problem is that I lack the vocabulary."[3]

Aristotle's *Energeia*

Aristotle first coined the term *energeia* from which the modern term "energy" is derived, although many modern philosophers and scientists see no relationship between the Aristotelian context and the modern meaning of the word.[4] In most translations of Aristotelian text, *energeia* is equated with the term "actuality." In *The Metaphysics* (Book IX, Theta) Aristotle noted the importance of *energeia* in relation to *dynamis* (potentiality) and *kinesis* (change or motion). He distinguishes two types of *energeia.* The first type of *energeia* is an unended or imperfect actuality, for example, walking or building. These *energeiai* are said to be movements or *kinesis,* because they are incomplete. The second type of *energeia* is a complete or perfect actuality, an *entelecheia,* where the end product is within the thing itself, for example, seeing and thinking. These *energeiai* are referred to in general as actualities.

In *Physics,* Aristotle elaborates upon his idea of change (movement) as a natural, continual process which is indefinite. Actuality (*energeia*) is dependent upon a thing's potential (*dynamis*) to do something. This "thing" is in turn contingent upon its ability to change. In other words, change (*kinesis*) is also, in a sense, an *energeia.*[5]

Aristotle believed that all potentialities arose from some sort of actuality: "We have distinguished the various senses of 'prior,' and it is clear that actuality is prior to potentiality."[6] To account for the continual source of change, he proposed that there was an unmoved mover, an eternal source of change, which was not dependent on anything else for its existence. It was pure actuality, *energeia,* although he claimed it to be a transcendent immaterial form. This notion of an unmoved mover was the impetus for later arguments concerning the existence of God.

Aristotle used four causes (the causes of *kinesis*) to explain the changing nature of all things. In direct contradiction to Parmenides, who argued that being was unchangeable and imperishable, Aristotle held that nature is essentially an inner principle of change: "nature... is the source of the movement of natural objects, being present in them somehow, either potentially or actually."[7] What he does is provide an explanation for why there is a tendency on the part of each thing to actualize into whatever it becomes. So strong is Aristotle's emphasis on the teleological qualities of things that there can be *no explanation for nature without it.*

Any "thing" is made of some stuff; this stuff without form is called the "material cause."[8] In this stuff a change can occur. The pattern or form into which this stuff can be changed is called the "formal cause." The "efficient cause" is simply the thing that brings about change or initiates the activity whereby the change takes place. Here we have an allusion to an idea of force as the thing that initiates change. The purpose or reason for which the change takes place, what Aristotle referred to as the "that for the sake of which," is called the "final cause."

A classic example is an oak tree. The germ of the oak, the acorn, is the material cause. The acorn, if it grows, will grow according to its own innate nature. It won't become a maple tree; if it becomes a tree, it will always become an oak tree. This is the formal cause. The efficient cause is the parent oak from which the new tree will grow. And, the final cause, the "that for the sake of which," is the new oak tree itself, having become what it now is. Consequently, everything in nature has a purpose. Unfortunately, with the advancement of science the word "cause" is usually referred to only in the context of Aristotle's *efficient* cause and the teleological implications have often been forgotten.

The *Science Times* recently ran an article about redwoods and fog.[9] Journalist Carol Yoon reported research findings which indicate that redwood trees transform fog into ground water and maybe stream water as well. The idea of fog contributing to our water source and plant life has been intuitively understood by ecologists for a long time. Apparently the amount of ground water that the redwoods transform from fog is so immense that the rest of the forest thrives because of it.

To put it all into an Aristotelian context, the material cause of the redwood is the seed from which it grows. The redwood seed will always actualize into a redwood, i.e., it will never become an oak — the formal cause. The efficient cause is the parent redwood, and the final cause is the new redwood tree that comes to be. But there are other questions we can ask about the redwood tree. For instance, why does it often grow so much larger than other trees? There seems to be a relation between the size of the redwood and its ability to transform fog into ground water.

Scientists hypothesize that the redwood is the tallest tree on earth *because* it is able to take fog and transform it into a great deal of ground water for its own growth and development. But then again, perhaps the redwood grows so tall so that it *can* transform more fog into ground and stream water in order for other forest life to maintain well-being. It would appear that the larger the redwood, the more fog it can transform into water. So it's possible that the redwood's ultimate purpose is to grow as large as it does so that it can transform more fog into ground water in order for other things to be able to actualize their potentials. And, perhaps stretching this a bit, maybe the ultimate final cause for *fog* is just so that redwoods and other plant life can transform it into ground water to begin with. Will we ever really know? One thing is for sure, the redwood tree has a purpose, the fog has a purpose, and together they are actualizing potentials.

There are two important antitheses prevalent in Aristotle's philosophy which are inextricably linked with his doctrine of four causes: matter-form and actuality-potentiality. When Aristotle wrote about what it meant for something "to be," he used the antithesis of matter-form, contrasting the unfinished thing with its final condition. When he wrote about the process of becoming something (coming into being), he used the antithesis of actuality-potentiality.

Matter is the stuff or material of which something is composed.[10] For Aristotle, matter when formed is usually observable, tangible, occupies space, and has the potential to change. Change can come about by the matter being acted upon by some other substance (efficient causes), or by way of exhibiting activity. On the other hand, form is the essence (*ousia*) in a particular thing. Form does not exist separate from matter; it is within the matter. This is where Aristotle deviated from Plato. Plato believed that the forms were separate, in other words, forms exist independently of any "thing." Aristotle held that forms could only exist independently of any "thing" in an abstract sense. Only in real things do forms actually exist. Further, it is the essence of a thing, manifested by its form, that makes it known to us as a particular thing.

Although the form is the essence of a thing and exists as a realized state, an *entelecheia,* it is also considered to be the dynamic force that propels the matter toward its final state. For instance, the seed in an apple is the form of the apple and also the reason why an apple ever comes to be at all. Aristotle's term *entelecheia* is synonymous with end state, or complete actuality, so that matter, when united with form, is both an actuality (an *energeia*) and an actualizing of potentials — an inner principle of change.

Philosopher Jonathan Lear uses Aristotle's example of the growth process of organic forms to express this idea:

> Form also provides the link between the mature and the immature organism.... For the form is and has been its nature throughout its development. The form is *both* that toward which the process is directed — "that for the sake of which" the growth occurs — *and* that which is directing the process.[11]

Thus, forms are irreducible, basic things that lend orderliness and purpose to matter.

Combining Aristotle's two antitheses, matter can be said to be a potentiality, and its form is, in essence, the actuality. For him, matter only comes to be potentially because it has a goal, an end, an actuality toward which it moves or changes. So all things in nature, and in art (all things created by humans) as well, progressed to the actualization of that which previously existed potentially *only* because of some inherent purpose:

> ...because everything that comes to be moves towards a principle, i.e. an end. For that for the sake of which a thing is, is its principle, and the becoming is for the sake of the end, and it is for the sake of this that the potentiality is acquired.[12]

Matter and the Forces of Nature

From the Grecian era up to the nineteenth century the term *energeia* disappeared, and instead the word "force" came to the forefront. As it was most often used, "force" clearly emphasized efficient causes, that which causes change in something else. Yet, imbedded in this idea of force is Aristotle's looser conception of *energeia* as change (*kinesis*). So that, not surprisingly, when the word "energy" appeared in the scientific literature it implied something different from *energeia* as an actuality. But as will be illustrated, the idea of energy as a thing — an actuality — eventually returned and, with it, profound implications.

Averroes (1126–98 A.D.), an Islamic philosopher and Aristotelian commentator, tried to clarify the constant motion of the heavens in accordance with Aristotle's belief that all *energeia* is caused to become, move, or change by something else.[13] He refined Aristotle's thesis by noting two types of complementary forces, a primary motive force and the secondary forces. Averroes proposed that the heavenly bodies drew an unending supply of force from the primary motive force, while conversely, the secondary forces replenished their quantity of force from the primary force. Such as it was, Averroes believed that the primary force was infinite in nature.

Judaist theologian Maimonides (1135–1204 A.D.), a contemporary of Averroes, also argued for the existence of God on the Aristotelian basis of an unmoved mover. However, unlike Aristotle, Maimonides believed not only that God was the efficient cause of all movement, but that God was also the formal and final cause for all things.[14]

Christian theologian Thomas Aquinas (1225?–74 A.D.) further developed the idea of force as an independent moving agent. Building on the work of his predecessors, he went on to form his famous argument for the existence of God, "The Five Ways." Aquinas started his argument with the thesis that because all things in the world are constantly changing or moving there must be a first cause or mover (the first of the Five Ways). He asserted that things move only because they are moved by something else. This being the case, he argued, the movements "cannot go on to infinity, because then there would be no first mover, and, consequently no other mover.... Therefore it is necessary to arrive at a first mover, moved by no other; and this everyone understands to be God."[15]

From the time of Aquinas on, philosophers and scientists were concerned with what they called the forces of nature. They came to recognize that certain *forces* seemed to always be conserved. However, there was some debate over just *what* was being conserved. Two conflicting views developed, in part laying the foundation for mechanism and dynamism. Instrumental in the development of these conflicting views were the philosophers René Descartes (1596–1650) and Gottfried Leibniz (1646–1716). Descartes said that momentum was always conserved, while Leibniz later argued that *vis viva*, or "living force," was always conserved.

Although Descartes is generally credited with the development of mechanism, mechanistic thinking is evident in some of the ancient Greek philosophies. It is based on a few fundamental assumptions, the first being that causal relations, specifically efficient causes, can explain things in nature. Every thing can be likened to a machine and can therefore be explained according to mechanical causes and the examination of parts. The second assumption is that there are no final causes, no "that for the sake of which"; everything occurs without intention or purpose. And lastly, according to this philosophy, all phenomena are physical, material things.

Descartes believed that all matter was simply extension, meaning that only one tangible substance constituted the entire universe. So here the idea of an essential oneness somewhat vaguely re-emerged. But Descartes noted: *"All the variety in matter, or all the diversity of its forms, depends on motion."*[16] In other words, all things are composed of material parts that interact through movement and are caused to change accordingly. Everything is similar to a machine reliant upon mathematical laws. From his initial theory of matter Descartes derived the notion that motion is always conserved. And finally, attributing the cause of this conservation to a deity, he too, like others before him, proposed *"that God is the First Cause of movement and that He always preserves an equal amount of movement in the universe."*[17]

Sir Isaac Newton (1642–1727) built upon the ideas of his predecessors, especially Descartes and Galileo, by formulating three laws of motion along with a concise mathematical delineation of the concept of force.[18] Force came to be understood as something that affects the movement of something else. In addition, the velocity with which something moves is dependent upon the strength of the force that causes it to move. In general, we can equate Newton's conception of force with Aristotle's efficient cause — the thing that brings about change or initiates the activity whereby a change takes place.

Newton's theories placed mechanistic philosophy firmly into the scientific arena. For everything was believed to be dependent upon an efficient cause and its effect. As such, the universe according to Newtonian science was a place devoid of purpose, dependent upon, and determined by, cause-and-effect sequences of a mechanical type.

Antithetically, Leibniz offered a different view from Descartes, arguing that a living force (*vis viva*) was always conserved. Interestingly, he regarded the most simple, indivisible units of matter, which he called monads, as the fundamental substances of the universe. Monads contained within themselves a kind of fixed program; for that reason, nothing outside could affect their being. Each monad was unique: "For in nature there are never two beings which are perfectly alike and in which it is not possible to find an internal difference, or at least a difference founded upon an intrinsic quality."[19]

Leibniz argued that the intrinsic quality in each fundamental substance is due to an innate force within the monad itself that directs its progress or activity (similar to Lucretius's fixed seed which accounts for changes in nature). Therefore, the continual changes that take place within the monad are caused by this innate force, which is always conserved. He likened these simple substances to entelechies: "for they have in them a certain perfection; they have a certain self-sufficiency which makes them the sources of their internal activities and so to speak, incorporeal automata."[20] Although the word "entelechy" is apparently taken from Aristotle's *entelecheia,* Leibniz used the word to imply an innate tendency in things, while *entelecheia* as used by Aristotle signified the final actuality or end-state of something.[21] Leibniz's entelechy is actually similar to Aristotle's final cause — the inner principle of change.

The argument raged on between the Cartesians and the Leibnizians, and although it was eventually acknowledged that the contentions regarding the conservation of momentum and living force were both in their own ways correct, two paradigmatic views seem to have emerged. These two conflicting views, that of a mechanistic versus that of a dynamic universe, not only continue to dominate all scientific disciplines; they unconsciously pervade our daily living and thinking. But for the most part, we have been conditioned by our education to see things as separate and particulate rather than as interwoven and inseparable aspects of a whole. We have been taught to believe in a Newtonian universe.

The Conservation of Energy

Scientist Johann Bernoulli (1667–1748) has been credited with the likelihood of being the first person to use the term "energy" in connection with mechanical work.[22] At this point if energy was mentioned at all it was usually regarded as part of a process in relation to motion and synonymous with force. Yet, the idea of forms of energy, which could be transformed into other forms of energy, was slowly emerging.

Studies of heat, biological, and chemical processes led to an understanding of the transformational nature of energy. Ludwig Colding, James Prescott Joule, and Julius Robert Mayer, scientists in Denmark, England, and Germany, respectively, among many others, deserve credit for enunciating the conservation of energy doctrine, albeit at slightly different times and in varying degrees. And although Mayer is often credited with the initial announcement of the conservation of energy doctrine, it has been recognized that here lies an exquisite example of simultaneous discovery.[23]

Mayer recognized a direct relationship between what was considered force (energy) and matter; to him both were causes in nature:

> In Nature we find two classes of causes. . . . The one class is made up of those causes which have the properties of ponderability (weight) and impenetrability, namely what we commonly call *matter*. The other is made up of those causes which do not possess these properties — these are the forces, which from the designated negative properties are also called *imponderables*. Accordingly forces are indestructible, transformable and imponderable entities.[24]

Thus, in 1842, from his observations and quantitative studies in biology and physiology, he concluded that there were various forces (forms of energy) that are interconvertible.

Joule came to the same conclusion during his studies of heat and mechanical power: "It is obvious that the force expended in setting a body in motion is carried by the body itself, and exists with it and in it, throughout the whole course of its motion. This force possessed by moving bodies is termed by mechanical philosophers *vis viva,* or *living force.*"[25]

But it was Colding who seems to have captured the teleological implications:

> The first idea I conceived on the relationship between the forces of nature was the following. As the forces of nature are something spiritual and immaterial, entities whereof we are cognizant only by their mastery over nature, these entities must of course be very superior to everything material in the world; and as it is obvious that it is through them only that the wisdom we perceive and admire in nature expresses itself, these powers must evidently be in relationship to the spiritual, immaterial, and intellectual power itself that guides nature in its progress; but if such is the case, it is consequently quite impossible to conceive of these forces as anything naturally mortal or perishable. Surely, therefore, the forces ought to be regarded as absolutely imperishable.[26]

Colding alluded to a purposiveness in what he called the "forces of nature." Moreover, like so many others before him who believed in a primary force or unmoved mover, he implied that the forces arose from a higher, supernatural power and that it was this power that gave purpose to every thing.

The principle of energy conservation really expresses ideas that sages have been forever writing about. There is some *one* thing underlying all things. As such it is a very *extraordinary* thing, a phenome*non,* because although energy exists autonomously and it cannot be created or destroyed, it can and does manifest itself in a multiplicity of ways. And, as scientists identified various energy forms and noted their interconvertibility, it became obvious to many that there is a teleological nature underlying the unity of all things. For many individuals believed that if the forms of energy are continually transforming although energy is never lost or created, then it stands to reason that the activity is likely to be purposive. Again, Colding's passionate words reflect this belief:

> Who is not aware that all the other heavenly bodies transmit light and heat to us, and that the Earth itself transmits light and

heat to all the other heavenly bodies in the universe? None of this is new to us; we all realize it perfectly well, and thus nature itself compels us to realize that the forces of nature, far from being bound to matter, are not even bound to the earth, but in fact, traverse the universe without cessation and bring light and clearness to every globe in the endless void, so that it may be clear to all that the goal of existence is the evolution of an ever more perfect spiritual and harmonious rational whole.[27]

Fields Everywhere

The nineteenth-century idea of a universe filled with force fields arising from the discovery of electromagnetic phenomena added more substance to the belief that unity and purpose underlie all things. Although most scientific models up until this time were largely dependent upon Newtonian mechanics, with the development of electromagnetic field theory the universe came to be regarded as embodying a dynamic nature. Matter, in the form of particles, was no longer considered the basic thing; instead, fields were elevated to the status of the most fundamental entities in nature.

Underlying the distinction between Newtonian mechanics and the dynamic idea of force fields there was, and still is, the conflict of whether or not things come to be because of some inherent purpose. For, as long as one espouses that everything has an observable cause and effect — a mechanistic belief — then one must disclaim a belief in final causes. Whereas conversely, a dynamic view often entailed the idea of forces in nature assuming some sort of directionality.

To account for electric and magnetic phenomena, Michael Faraday (1791–1867) intuitively conceived the idea of imaginary lines of force. The purpose of these imagined lines of force was to lend some kind of visual direction to the forces, and not the material things themselves.[28] Faraday oscillated between the idea of these lines of force being contiguous (next to or joining one another as conductors) and the notion that they were continuous (inseparable, physically real forces pervading space).[29] So once again we have an idea of separate, distinct things acting on one another (efficient cause) versus the idea of *one thing* acting as a whole.

But Faraday also experimented with chemical affinity, heat, light, magnetism, and electricity and consequently noted their ability to convert one into another. Having completed a series of such experiments with students he stated:

> I might show you many other experiments whereby I could obtain electricity and chemical action, heat and light from a magnet, but what more need I show you to prove the universal correlation of the physical forces of matter and their mutual conversion one into another?[30]

Indeed the *forces of matter* can be equated with various forms or manifestations *of one thing, energy.*

Faraday's conception of force fields formed the basis for James Clerk Maxwell (1831–79) to develop the electromagnetic field theory. The ideas underlying field theory are based on a number of formulas called Maxwell's equations, a mathematical language familiar only to certain scientific disciplines, and as such these ideas have been for the most part beyond the layperson's comprehension.

A classical physicist once told me that "Maxwell's equations say everything there is to know about fields, and that is all there is." But Maxwell's work opened up a whole new world for scientists, and I couldn't help feeling that there must be more to field theory than mathematical formulas that most of us, myself included, don't understand.

The general gist of Maxwell's equations is that they describe the electromagnetic field itself, aside from the things (wave-particles) in it.[31] In other words, the field *is* a distinct *thing*. Maxwell noted that electromagnetic field theory seems to rest on the idea of energy residing everywhere:

> In speaking of the Energy of the field . . . the only question is, Where does it reside? On the old theories it resides in the electrified bodies, conducting circuits, and magnets, in the form of an unknown quality called potential energy, or the power of producing certain effects at a distance. On our theory it resides in the electromagnetic field, in the space surrounding the electrified and magnetic bodies, as well as in those bodies themselves.[32]

So, essentially a field is an area that may or may not have *discernible* matter in it; it may even be perceived as being empty. But we now know better: nothing is empty. In this empty looking space, or field, it is believed that things are interacting, possibly moving, but definitely changing because of the continually changing manifestations of energy in the field. Further, it is really not just one field, but a multiplicity of fields which are all overlapping one another. Following behind the enunciation of the conservation of energy, field theory verified and expanded the idea of energy as an autonomous entity.

The field is said to be a large store of energy, and it has the potential to produce a force. There are multiple fields in space, and they all create forces that relate to each other. In electromagnetic field theory, energy is the invariant entity; the total amount of energy among all of these fields does not vary even though we may not be able to see or measure it directly. Of course, *nothing stands still.* As J. H. Poynting, another scientist in Maxwell's day, expressed it, the energy is always present, always flowing:

> If we believe in the continuity of the motion of energy, that is, if we believe that when it disappears at one point and reappears at another it must have passed through the intervening space, we are forced to conclude that the surrounding medium contains at least a part of the energy, and that it is capable of transferring it from point to point.[33]

The recognition of the independent nature of energy, for no longer was it necessary that matter be the vehicle for its continual transformations, led to the belief that energy could, by itself, be considered a substance. From this discovery arose the philosophic doctrine of energetics, which holds that energy is the fundamental entity from which all else is derived.

The Tiny Quantum World

> There is an old saying about the reception of a new idea, which goes as follows. At first it is dismissed as being wrong; then it is characterized as being against religion; and finally it is said to be something everyone knew all along. — R. LEWEN[34]

Modern physicist Richard Feynman had written: "Energy has a large number of different forms.... These are: gravitational energy, kinetic energy, heat energy, elastic energy, electrical energy, chemical energy, radiant energy, nuclear energy, mass energy."[35] What this basically implies is that *we* are energy, as are trees, birds and bees, puppies and kittens, hamburgers, hot dogs and ice cream too. It means that Bach's sonatas, Michelangelo's Sistine Chapel, da Vinci's *Mona Lisa,* Dante's *Divine Comedy, everything* is in essence a manifestation of *one* thing.

Einstein's publication of the special theory of relativity provided the impetus for us to acknowledge that matter and energy are interchangeable. The word "relativity" in the relativity theories can be misleading to the casual reader. Einstein did not use the word "relativity" to imply that all things are relative, but to emphasize the fact that we, the observers of everything around us, are participants *in everything around us.* Theorizing that the observer's place and motion play a part in what is observed, Einstein eventually came to believe that the laws of nature should be the same for everyone despite individual placement and motion. However, because we as observers play a part in what is observed, to a certain extent we create reality.

Einstein's formula $E=mc^2$, first published in 1905 (E is the energy of a system, m is the mass, and c is the velocity of light), implies that energy and mass are interchangeable. Although he never really said energy and mass were the same, from his view came the belief that the universe is a large web of energy rather than a Newtonian machine with various parts. Einstein's theory basically says that all mass is equivalent to energy, but it also implies that all energy is equivalent to mass. That would mean that mass-less things are not equal to energy, which in turn means that photons (light particles), which are mass-less, and other similar particles (i.e., for now, the virtual ones) are not energy. But physicists have gotten around that with the idea of kinetic energy — the energy something has when moving. When something moves, it apparently gains mass. So mass-less particles are said to have zero rest mass, which essentially means that although they have no mass, they *have* energy — kinetic energy.

All the "stuff" composing material things can be transformed into other forms of energy — which reminds me of the movie *Phenomenon.*[36] I really like the name, which shows that someone grasps the fact that energy is a phenome*non* and not one of many phenomena. It is, after all, one thing, albeit many forms. The movie is all about energy, particularly human energy and the inherent potential all of us have to go beyond the superficial physical realm. But I am jumping ahead here; human energy is a topic for other chapters. One thing is for sure though: we are certainly moving into a different era if the new motion pictures and books are any indication. There seems to be a greater awareness of an essential oneness than ever before. In this modern view is the belief that no single entity or part can be affected without all the other parts being similarly affected. Everything in this universe is inextricably interwoven into one great association. Even space and time are indivisible and form a multidimensional continuum, space-time.

The leap from reality being absolute to being relative surely makes sense when one considers the intangible nature of many forms of energy. Take physicist Wolfgang Pauli's elusive neutrino. Back in the 1930s he had hypothesized the existence of a particle that could account for energy that seemed to be missing when neutrons decay.[37] For decades scientists believed that these particles were mass-less. Today it is known that there are three kinds of neutrinos (electron-neutrinos, muon-neutrinos, and tau-neutrinos), and it looks like they have a tiny bit of mass after all. Science writer Dennis Normile notes that the recent data is significant enough to "provide strong evidence that a certain type of neutrino from the atmosphere is "disappearing" by changing, or oscillating, into another type of neutrino the detector can't see."[38] Only if the neutrino has mass could this change occur. What they are saying is that one form of energy is transforming into another form we can't, as of yet, see.

Yes, there are still a lot of gray areas in physics. In a 1994 issue of *The Sciences* there was a catchy cover article "Beyond $E=mc^2$: A First Glimpse of a Universe without Mass." The scientists, B. Haisch, A. Rueda, and H. E. Puthoff, posited a dynamic theory in which "mass, inertia and gravity arise from underlying electromagnetic processes."[39] I will not attempt to simplify their complex theory; however, they note that it leads to the possibility that "there is no such thing as mass — only electric charge and energy, which together create the illusion of mass."[40] Thus, these scientists interpret $E=mc^2$ as "Mass is energy."[41] This bold reinterpretation leads to the supposition that there is really no such thing as mass and consequently inertia is nonexistent as well. What is happening instead, claim Haisch and colleagues, is that electromagnetic forces are somehow acting on the electric charge inside the matter creating the effect of inertia.

Inertia is said to be the property of matter that keeps it in its normal state of rest or uniform motion, unless or until an external force comes along to change the material things' normal state. The term "inert," from which "inertia" comes, implies motionlessness or inactivity. But does anything in this universe stand still? In mechanics, where the language evolves around forces causing effects, inertia might make sense, but in the everyday world of both the microcosm and macrocosm, it has little appeal, once one comes to terms with the fact that *nothing stands still.* From moment to moment *everything* is changing; nothing is ever the same. Just as a river changes with each passing moment so that it is always new. Just as the sky, although always the sky, is never the *same* sky from day to day or moment to moment. Continual change is everywhere evident.

The Paradoxes of Nature: Matter/Ether, Particles/Waves, Local/Nonlocal

Not all forms of energy are tangible, and so we don't think about them as things. The conceptualization of heat, light, and the ether are notorious examples of this problem. Heat, from early on in antiquity until the nineteenth century, was a source of scientific discussion and debate. The problem of heat had its basis in its classification as a tangible thing (the atomic, or caloric, theory of heat) versus the idea of an intangible entity arising from a process (the vibration theory of heat). Light is another energy form that isn't really tangible. Newton had established, or so everyone thought, that light is composed of corpuscles of matter; however, the notion of fields and various experiments led to the conclusion that light is composed of waves. Once again there was a consistent difference between the two theories, the difference between a tangible thing, namely, particles of light, and an intangible thing, namely, waves of light.

Another marvelous example of our difficulty with conceptualizing intangible things lies in the ancient and somewhat still pervading notion of an elusive incorporeal element called ether. The ancient Greek philosopher Anaxagoras had used the term *aether* to express an elusive type of air that he believed was dispersed throughout the universe. Aristotle then used the notion of *aether* (quintessence)—an unearthly, immaterial thing—to account for the heavens.[42] Democritus, discounting the idea of a void in his atomist theory, also supported the idea of an *aether* and attributed the movement of the heavenly bodies to this medium.

Centuries later, Newton and others reintroduced the idea of ether to try to account for the theory of light waves and the force of gravity in space. This time it solved a scientific dilemma: the ether became a medium that pervaded everything so it could transmit light from one place to another, and it could account for the seeming actions-at-a-distance that no one could explain. So where one might suppose there was nothing, it was believed there was this subtle intangible form of matter.[43]

Up until the early 1900s, the ether was the only way scientists could account for light transmission and action-act-a-distance phenomena. Sir Oliver Lodge, a renowned British physicist, spoke of the ether as the foundation for all life energy sources. Yet, he also noted the elusiveness of this ether. Seeing a direct relationship of the ether to energy as the source of life he stated:

> We are entirely dependent on the ether, or what we call empty space, for our existence.... So in classifying forms of energy, it would be absurd to attend to matter alone and ignore the ether. For energy only enters matter in secondary fashion; it exists primarily in the ether, and then manifests itself by means of matter.[44]

Lodge continued to draw a distinction between the matter we believe to constitute life and the possibility that life may not exist only in matter:

> It is possible that life exists primarily in the ether, that it only enters into relation with matter in secondary fashion for a time, and then returns whence it came.... And, after all, it can only be an analogy until we begin to know what life really is. But just as it would be a mistake to suppose that energy could exist only in relation to matter, so also it may be a mistake to suppose that life only exists in relation with matter.[45]

This is a fascinating and radical line of thinking for someone born of the materialist-mechanist era. It is also interesting to note that Lodge participated in studies of paranormal phenomena.

After numerous attempts to empirically demonstrate the existence of an ether, the idea was again discarded in the early part of this century; Einstein and Infeld went so far as to say, "After such bad experiences, this is the moment to forget the ether completely and to try to never mention its name."[46] But the ether has recently been reintroduced once again to try to explain a mysterious aspect of the cosmos. So we have, at least for the time being, settled for a kind of matter-ether explanation for the cosmos. The ether saga will continue in the next chapter.

It seems we now have a tangible-intangible explanation for everything. Light and other forms of energy have a dual action: sometimes the energy behaves like particles and sometimes it behaves like waves. Taken one step further, all forms of energy are both particles and waves. This discovery was due in part to Max Planck's theory of the quantum of action in which he proposed that light traveled in discrete energy packets called quanta. This means that the light energy being emitted and absorbed was discontinuous and particle-like, which subsequently contradicted the wave theory of light. Planck noted the theoretical implications of his discovery:

> Perhaps there is no more impressive proof of the fact that a pure wave theory cannot satisfy the demands of modern Physics any more than a pure corpuscular theory. Both theories, in fact, represent extreme limiting cases. The corpuscular theory, which is the basis of classical mechanics, does justice to the configuration of a system, but fails to determine the values of its energy and momentum; conversely the wave theory, which is characteristic of classical electrodynamics, can give an account of energy and momentum, but excludes the idea of the localization of light-particles.[47]

What is really fascinating is that waves are not considered to be actualities; they are considered probabilities. Renowned physicist Werner Heisenberg referred to the probability wave as:

> ... a tendency for something. It was a quantitative version of the old concept of "potentia" in Aristotelian philosophy. It in-

> troduced something standing in the middle between the idea of an event and the actual event, a strange kind of physical reality just in the middle between possibility and reality.[48]

So here we are right back to Aristotelian concepts, and I wonder if that "tendency for something" is actually the final cause.

Heisenberg discovered that the position and the velocity of an electron cannot be measured simultaneously with any degree of accuracy, which formed the basis for the uncertainty principle:

> One could speak of the position and of the velocity of an electron as in Newtonian mechanics and one could observe and measure these quantities. But one could not fix both quantities simultaneously with an arbitrarily high accuracy. . . . Similar relations could be formulated for other experimental situations. They are usually called relations of uncertainty. . . . One had learned that the old concepts fit nature only inaccurately.[49]

Consequently, we cannot measure various things on a quantum or microscopic level with any degree of accuracy unless we take into consideration the measuring device and the other experimental conditions, even the observer, around the particular thing we are trying to measure. The supposition here is that parts cannot be explained without taking into consideration the whole environment.

So now we have another dilemma. How can we, the conscious observers, participate in creating change with something that doesn't exist? If waves are only probabilities and not actualities, how can we *cause effects* by observing nothing? So we must think of an energy wave function as being the same as an energy particle; they are two aspects of one reality, a reality that we participate in creating.

It was but one small leap from the uncertainty principle and the dual wave-particle character of matter to physicist Niels Bohr's theory of complementarity.[50] He proposed that on the quantum level nothing can be divided into discrete parts. Everything is related; everything complements everything else. Since it is impossible to completely predict outcomes on a quantum level, we are forced to look at the whole. Furthermore, Bohr illuminated the philosophical issues entwining complementarity in the quantum physics world with psychic experience and the study of living organisms in general. The simple fact is that everything is inseparably connected.

The theory of nonlocality, first proposed by Einstein, Podolsky, and Rosen in the 1930s and later further investigated by Bell in 1964, added even more credence to the notion that everything is inseparable. A particularly interesting experiment (generally called EPR experiments after Einstein, Podolsky, and Rosen) performed to test Bell's theorem showed that paired photons (light wave-particles), when split, remained somehow interrelated despite the conditions imposed upon them by the experiment. In other words, the twin photons seemed to *know* what was happening to one another. Even more recently, journalist George Johnson reported that researchers in Geneva confirmed and extended the finding that "two photons separated across great distances remain 'correlated' in some funny mathematical way that defies common sense."[51] Scientists John Briggs and F. David Peat liken EPR experiments to a story about a pair of Siamese twins who, while separated at birth, still experienced each other's feelings of pleasure and pain even though they lived far away from each other.

The principles of uncertainty, nonlocality, and complementarity, are not limited to quantum physics. They are significant for many scientific disciplines, particularly the health sciences, where we've been busy trying to understand parts in relation to the whole but haven't been completely successful. And that's because there are no parts; there is only a whole.

These far-reaching ideas continue to be substantiated by the more recent research in quantum physics. For a while now physicists have known that an atom can be in more than one place at a time.[52] However, recently they have demonstrated that not only can an atom be in two places simultaneously, but also that the distance by which the atom is manifested in two places at the same time is so large, in quantum terms, as to possibly be applicable on a larger scale.[53] Every day the possibility grows ever stronger that there are links between the microcosmic universe of quantum physics and the macrocosmic universe in which we live.

The emphasis by these scientists has been placed on the fairly recent discovery of the *distance* between which the atom is said to exist in two places. However, perhaps we should place the emphasis on the certainty that the atom *can* be in two places at once. The fact that an atom, a specific thing, a distinct energy form, can be in two places at once leads to the very probability that it can exist simultaneously in even more than two places, maybe even everywhere! Accordingly, perhaps we can speculate that since the atom doesn't *seem* to exist everywhere at once, *or at least we can't perceive that it does,* there is a possibility that the atom has *chosen to manifest* itself in a specific place, at a specific time.

So let's recapitulate the ramifications of all this for the enigma of energy. We now have energy, specifically manifestations of energy on a quantum level, which are unpredictable. They are unpredictable because when one tries to measure or observe the energy the characteristics or outcome of the energy (as a system) changes. The measuring apparatus and the observer essentially become part of the energy system, that is to say, there is a mutual or reciprocal action between the quantum energy forms and the environment.

Now, it may be oversimplifying things a bit, but it does seem plausible that we can't measure the quantum energy form or its movement simultaneously, because *all is essentially one.* Briggs and Peat surmised the evident paradoxical nature of quantum theory noting:

> All the cherished ideas of physics and everyday common sense had been challenged by quantum behavior. Light is both wave and particle, electrons are both particles and waves. When a particle is shot from one point to another, it possesses no path in between. *Where* is the electron during a quantum jump? *How* can something be both a wave and a particle?[54]

It seems that the electron possesses no path in between, because it is the same thing as everything else around it. It is all energy, so it exists *everywhere* at the same time, although it may choose to manifest itself in a specific form, in a certain place, at a certain time.

Summary

With Aristotle's various uses of the word, we come to two related and yet different assumptions concerning *energeia. Energeiai* are actual, complete things (*entelecheia*) and at the same time, *energeia* is also the process by which a potential thing comes to be actualized (*kinesis*). Overall, Aristotle's ideas are often referred to as animistic, and as Paul Davies, a noteworthy physicist, notes: "There exists, he [Aristotle] maintained, what we might today term a *cosmic blueprint*."[55] The teleological implications in Aristotle's work continued to be a predominant theme in the conceptualization of energy.

Those who came after Aristotle drew upon and further developed not the idea of *energeia* (actual entities), but the idea of forces or efficient causes. However, along the way, in identifying and coming to understand the purported forces of nature, there was a return of interest to the changing nature inherent in things, separate from the cause(s) of the change. At this point the term "energy" became a distinct reference to activity/motion or, in a more general sense, to the changes occurring in some thing. But use of the word "energy" did not just imply actual activity or change, it also expressed the ability of some *thing* to perform activity or to create change, whether or not the activity or change was actually taking place.

It seems that if the forces were to be treated as inherent in things (Leibniz's contention), the relation between Aristotle's philosophy and this dynamic view is distinct. For it would mean that the potential for change is within things, rather than outside of them. This implies the possible existence of both formal and final causes, and is certainly congruent with Aristotle's philosophy. Conversely, if we assume that in order for something to change there must be a force exerted upon it, we have the idea of efficient causation. This idea is the philosophical underpinning for mechanism. For example, in Newton's view the forces appear to come from without rather than from within.

Clearly profound philosophical implications came to the forefront with the enunciation of the doctrine of energy conservation, namely, the autonomous and seemingly teleological nature of the phenome*non* of energy. Yet scientists still advanced the idea of energy as part of a process. Energy thus expressed refers to the ability to do work or activity rather than some *thing* that is manifesting itself in a multiplicity of ways, something that is unceasingly transforming or changing its nature.

As scientists have moved deeper into their explorations they have found that, contrary to Newtonian physics, there are no separate entities. Every day new discoveries point to an incredibly organized and integrated complexity in our universe. I imagine it is scientific discoveries like the ones herein described that have motivated the creation of motion pictures like *Powder*[56] and *Phenomenon* and inspirational novels like James Redfield's *The Celestine Prophecy.* All of these creative renderings are basically about the phenome*non* of energy.

Relativity theory along with the principles of uncertainty, complementarity, and nonlocality provide strong evidence for the argument that all is essentially one — an undivided whole. It does seem that the innumerable manifestations of energy are indivisible, and as such it is likely that we will never be able to fully comprehend, predict, or explain everything in existence. Yet, there have been notable attempts to theoretically explain the underlying order found in nature using the idea of hidden variables. In the next chapter we will explore some of these ideas.

Can the modern scientific discoveries be reconciled with Aristotle's idea of *energeia* as actuality? Yes. Energy is an actuality, a thing. But more importantly, it is *one thing.* Science and philosophy are finally merging their beliefs as nature is being increasingly recognized as dynamic rather than mechanistic. Underlying this emerging view is the idea of a seemingly creative tendency in the universe to actualize some potentials rather than others.

All things have a definite rhythmical pattern that is constantly changing. Activity that is probabilistic but not predictable is innate in all nature. Even on a quantum level things are predictably unpredictable! Movement is in a definite direction toward something not yet actualized. And the direction a thing moves in is for the benefit of its own becoming. Therefore, the modern view of nature and hence, our universe, can be equated with principles set forth by Aristotle centuries ago.

In the next chapter we will venture into the larger realm of "becoming." The larger realm of becoming entails the evolution of the universe and life, in all of its manifestations. It is the domain nearest to us — the macroscopic domain that we observe and experience every day. And perhaps it is at this level of becoming that the idea of energy becomes most profoundly baffling.

Chapter 3

The Becoming of It All

> Something buried deep in the human psyche compels us to contemplate creation. It is obvious even at a casual glance that the universe is remarkably ordered on all scales.... From whence came the myriads of galaxies, stars and planets, the crystals and clouds, the living organisms? How have they been arranged in such harmonious and ingenious interdependence? The cosmos, its awesome immensity, its rich diversity of forms, and above all its coherent unity, cannot be accepted simply as a brute fact.
>
> — PAUL DAVIES[1]

It is fair to say that nowhere does it become more apparent than in the proposed theory of the origin of the universe, fittingly called "the big bang," that the becoming of it all is a result of one phenome*non,* one thing — energy. Throughout the ages many theories have been proposed for the becoming of it all; however, with the advancement of science the big bang theory has been pretty much elevated to the status of the explanation for the origin of the universe. It is a partial one at best, for it does not completely eradicate the mystery underlying everything in existence. Of course, this mystery is linked irrevocably with that of the energy enigma. In *The Awakening Earth,* Peter Russell wrote:

> One of the first things we notice is that the Universe today contains characteristics of which there was no trace in the beginning. Immediately after the Big Bang, there was only energy. Out of this developed a whole new order of existence: physical matter. For eons, this matter was inanimate, yet out of it emerged a new order: life. Life persisted and flourished and from living organisms emerged another new order: self-reflective consciousness.[2]

Back to the Forces of Nature

Crucial to the development of theories relative to the becoming of it all are the four fundamental forces. The four forces are usually referred to as the fundamental *forces of nature,* and of course this is reminiscent of the historical period during which time the term "force" was used in lieu of "energy." Scientists have distinguished many different types of matter particles, which for convenience's sake, I refer to as the "material energy group." Within this material energy group there are force-carrying particles. These force-carrying particles are often referred to as *virtual* particles because they cannot be detected (as of 1999, anyway!) and they fit into one of four categories: gravitational, electromagnetic, weak nuclear, and strong nuclear forces.

Gravitation is responsible for binding stars into galaxies and for keeping the planets in their respective orbits around the sun. Physicists use a virtual particle, that is, a hypothetical particle called the graviton, to explain the force between two matter particles. Renowned physicist Stephen Hawking notes:

> The gravitational force between the sun and the earth is ascribed to the exchange of gravitons between the particles that make up these two bodies. Although the exchanged particles are virtual, they certainly do produce a measurable effect — they make the earth orbit the sun! Real gravitons make up what classical physicists would call gravitational waves, which are very weak — and so difficult to detect that they have never yet been observed.[3]

The electromagnetic force is responsible for keeping electrons surrounding the nuclei for the formation of atoms as well as binding the atoms together. The strong and weak nuclear forces also dominate the world of atoms; the strong force acts like an epoxy sticking protons and neutrons together to form atomic nuclei while the weak force brings about radioactive decay. Thus, all of these forces are, in a sense, responsible for the properties of matter.

The four known *forces of nature* seem to act simultaneously toward, at the very least, some mutually inherent goal. Whether or not there is an end product is not at issue here; rather, at issue is the fact that these forces together comprise some sort of self-organization. The difficulty physicists have had with the four forces of nature has been that while three of them — the electromagnetic, the strong, and the weak nuclear forces — fit together nicely and seem to account for events in the microcosm, the gravitational force is still an outlier, accounting for things in the macrocosm. To date the four forces appear to be incompatible or at least incapable of being unified into one theory of everything (TOE). The challenge, then, is for physicists to unite the four forces and, consequently, the two major theories — quantum physics and relativity — into one theory that can explain everything. One theory proposed that attempts to do just that is superstring theory.

It has been theorized that everything is composed of tiny connected strings. Of course there are a number of superstring theory models as well. The newest theory of everything, a derivative of superstrings, adds in membranes and black holes and is called M-theory. But in essence all of these theories really do support the idea that all is one, although the all perhaps consists of an infinite number of tiny strings resonating throughout the cosmos.

Big Bang(s)

There are a number of models of the big bang theory, and most of them incorporate the hypothesis that periods of rapid chaotic expansion (inflation) occurred in the early stages of the universe's evolution. The most recent inflationary model proposes that there may be multiple universes arising from an eternal inflationary process. This would mean that the universe is constantly re-creating itself, perhaps from multiple singularities rather than one, which implies, as cosmologist Andrei Linde notes, that "the universe as a whole is immortal."[4] In essence we could have multiple big bangs occurring. But Linde also notes that inflationary theory still does not answer the question of how it all got started: "The situation with the very beginning is less certain. There is a chance that all parts of the universe were created simultaneously in an initial big bang singularity."[5] So within the context of some kind of big bang we can further examine the enigma of energy.

Something's Missing

Some big bang models suggest that the universe will continue to expand forever, while others hold that eventually the universe will stop expanding. Still others had predicted that eventually the universe would collapse into a big crunch, but that theory hasn't been able to hold up with recent discoveries because it seems that not only is the universe expanding, the rate of expansion is apparently accelerating. One thing is clear. Cosmologists have not been able to reconcile the expanding universe with the amount of mysterious energy (or mass) that is *known* to exist. In other words, there appears to be some *thing* missing. Remember Pauli's neutrino? The fairly recent discovery that neutrinos have some mass resulted in a short-lived enthusiasm that these particles might be able to account for the missing mass. But scientists are now fairly certain that it's a tiny amount, simply not enough to provide whatever is missing.

According to various models of the big bang, there should be x amount of mass in the universe, but scientists can only account for only a tiny percent of the stuff. For quite a while now, this missing mass has been called "dark matter." Some cosmologists think it's baryonic matter—normal ordinary stuff (a.k.a., MACHOS, massive compact halo objects), like that which comprises everything we see, while others speak of exotic, *invisible* matter (a.k.a. WIMPS, weakly interacting massive particles).[6] So here we go again, trying to account for things of an intangible nature by positing some kind of material thing that is invisible. Now it seems that although scientists haven't exclusively given up the hunt for dark matter, they are also hunting for some kind of mysterious "force."[7] Scientists usually refer to the dark matter as "missing mass" and the mysterious force as "missing energy."

The mysterious *force* (a.k.a. missing energy) also goes by other names like "funny energy," "something strange," and the good old "cosmological constant," coined by Einstein to add into his field equations for general relativity theory in order to account for the seeming stability of the universe.[8] Although Einstein later discarded the "cosmological constant," it keeps cropping up again, as science writer Dennis Overbye notes: "the cosmological constant, like Rasputin, has stubbornly refused to die, dragging itself to the fore, whispering of deep enigmas and mysterious new forces in nature."[9]

The most interesting name by which the missing energy goes is that of "quintessence," which is reminiscent of Aristotle's element, *aether.* This missing energy, cosmological constant, ether, quintessence, or whatever else one chooses to call it, is thought to be some kind of antigravity force, a repulsive force that is speeding up cosmic expansion instead of slowing it down. Whatever scientists call it, it certainly does seem as if they have come full circle. Force is once again energy and the ether is back besides![10] Whatever the stuff is, we know that what appears to be a void of nothingness is not; it's filled with an unending foam of virtual particles that fluctuate in and out of existence and obviously produce measurable quantum activity. Out of this, somehow, came the becoming of it all.

The Universe Unfolds

Ten to twenty billion years ago a spark of energy is believed to have created a force so powerful that it resulted in this huge and, some say, infinite universe. The big bang theory holds that the universe resulted from an explosion of some primordial energy (or a virtual particle) and that ever since, our universe has been expanding. This simply means that our universe is still in the process of becoming, and it seems clear that it is becoming with increasing complexity.

Before the big bang the universe was in perpetuity contracted, meaning that there was no distance between any two points, the density of matter was infinite, and the volume of the entire universe was zero. To understand this idea, it might help to try to picture something so minuscule you cannot really envision it. The distance between two points equals zero, but the mass of the "thing," is unfathomable. So much so we could say it was boundless, like Anaximander's infinite, intangible substance *aperion.* Yet the volume of this thing also equals zero, so that you are actually seeing nothing, although something is there. The "thing" or primordial mass, simply put, was the phenome*non* energy. This primordial energy, although it appeared to be a void, was packed with enough power to form everything in our universe.

In the first 10^{-43} second (a lot less than an actual clock-time second) after the big bang, the universe was a big blaze and the heat was beyond our wildest imaginations. It is believed that as the explosion occurred, the primordial energy expanded and gradually cooled down, allowing the first elementary particles — quarks, electrons, and neutrinos — to form. These particles were teeming with photons (light particles). In his picturesque book *The Birth of the Universe,* Trinh Xuan Thuan notes how matter appears to have "froze out" of an energized void:

> When matter came into being, a mirror image, antimatter, was created, which counterbalanced its electrical charges and thus preserved the electric neutrality of the universe. Matter and radiation were constantly interacting. When a particle and its antiparticle collided, mass was converted into the energy of photons. Photons were in turn converted back into particle-antiparticle pairs. Matter, antimatter, and photons were continuously dissolving and reforming in infernal cycles of life and death.[11]

Strange it is, as physicist Paul Davies notes, that "matter, which is a form of energy, can be created in the laboratory, but it is always accompanied by an equivalent quantity of antimatter, yet our universe is composed of almost 100 percent matter."[12] It's almost as if energy *chose* not to manifest itself as antimatter. And it is really a good thing because matter and antimatter cancel themselves out: they annihilate one another.

Well, somehow matter managed to outnumber antimatter and our universe, with its multiplicity of energy forms, came into being. Cosmologist John Leslie indicates "that the initial cosmic speed of expansion had to be correct to about one part in ten billion billion billion billion billion billion to avoid the rapid recollapse of everything or else its quick dispersion into cold, enormously rarefied gases."[13] Some cosmologists, including Leslie, consider this partial evidence that our universe has been fine-tuned.

The apparent evidence for universal fine-tuning is called the anthropic cosmological principle and is based on three things: the rate the universe appears to be expanding, the particle-antiparticle ratio, and element formation.[14] Many individuals, such as scientists John Barrow and Frank Tipler, assert that when combined, these things offer a fairly strong argument for universal design and, ultimately, a designer. For, evidently, with just a tiny bit of a slip we would not be here.

From Stardust to Us

As the universe cooled, the first elementary particles, with the assistance of the fundamental *forces of nature,* clumped together forming hydrogen and helium nuclei and then, later on, atoms. Photons and other forms of radiation flew throughout space creating what today is referred to as microwave background radiation.[15] Then galaxies and, after a very long time (approximately two billion years), stars finally formed.

The saying that everything has come from stardust is not fictitious. By way of a very long, intricate process the stars in the universe began to create the chemicals which would later on provide the universe with life in all its wonderful aspects. Among all the stars there exists a type of star referred to as a supernova: an exploding massive star that has reached a critical point where it can no longer sustain itself.[16] And as geophysicist Robert Jastrow notes, with the explosion of supernovas the elements necessary for the future development of life were thrown chaotically about space. So essentially a supernova is a form of energy that changes or transforms into something else at a certain point. The something else in turn becomes something else, which then transforms into something else ad infinitum. The first supernovas apparently became something else, which became something else, and eventually down the line, this something else became *us*. Hence, our ancestors are supernovas; we are essentially composed of stardust.

How is it that this phenomenon, energy, has transformed itself into a multiplicity of meaningful forms? For it does seem that the continual transformation of energy is purposive. Even black holes are somehow another purposeful manifestation of energy. Hawking notes that black holes emit particles and in a sense play a part in the balancing act of energy. He gives this example:

> The existence of radiation from black holes seems to imply that gravitational collapse is not as final and irreversible as we once thought. If an astronaut falls into a black hole, its mass will increase, but eventually the energy equivalent of that extra mass will be returned to the universe in the form of radiation. Thus, in a sense, the astronaut will be "recycled."[17]

Hawking also notes that "the only feature of the astronaut that would survive would be his mass or energy." Now if one perceives the soul as another meaningful manifestation of energy, then the astronaut's soul consequently would survive. But we are jumping ahead; there will be more on the soul later.

The long, intricate process of change, which resulted in the universe we know today, is most often referred to as a number of energy processes. Specifically, terms such as "the generation of energy," "use of energy," "mass converting into energy," and "energy thermonuclear reactions" are used to depict the becoming of it all. Of course these terms only confuse the issue of the energy enigma; they imply that energy is some thing that exists only as part of a process. The energy is seen to be outside the thing rather than the thing itself.

However, as previously described, there are also those who have alluded to energy as some "thing" which exists and is by itself constantly transforming or changing. That would mean the changing nature is within the phenome*non* of energy itself. In other words, the primordial energy — the phenome*non* — is responsible for everything we see (and don't see) in the universe, including us. It is responsible for the becoming of it all.

Often the term "evolution" is used to denote the becoming of the universe and all we see around us, including us. Yet the word seems such an inadequate one for such a vast and beautiful becoming. That's partly because when many of us hear the word "evolution," we tend to think of Darwin's theory of natural selection along with the connotation that we just happen to be here. But in essence, all theories of evolution are theories explaining ongoing energy transformation.

While researching the topic of human evolution for this inquiry I happened upon a book written by C. Pellegrino and J. Stoff. In the preface Stoff wrote:

> Scientists and philosophers have long debated the question of what brought about the monumental transition from inanimate matter to living organism. My own studies lead me to believe that there is an "energy principle" that is basic to life. Life, in essence, is the creative energy that orders the entropic patterns of matter in accordance with its own rhythmic processes.[18]

It seems that there is an energy principle that is basic to life, for life is another manifestation of one phenome*non* — energy — which has actualized into an orderly pattern.

Approximately 3.5 to 4 billion years ago life began on this planet. The first life forms were single-celled microscopic things that are said to have evolved from prebiotic organic molecules.[19] This means that life appeared from a "nonliving" form of energy. It took approximately 15 billion years for humans to develop into our current form. And it took billions and billions of particles to come together in just the right way. The progression to human status took place during the last billion years or so, although human beings with the capacity for thinking evolved from approximately 1 million to 100,000 years ago. The inordinate length of time it took for this evolutionary process to result in human beings is part of the theory that Darwin articulated in his work *The Origin of Species.*

According to Darwin, the evolutionary process is one of trial and error. Darwin's principle of natural selection is based on the premise that all living things compete for the essential resources in the environment.[20] The fittest of the living things, the ones best equipped to get what they need from the environment, will survive. Hence, according to Darwin, there are no final causes in nature. Speaking of the morphological structure of living things he clearly denied final causation: "Nothing can be more hopeless than to attempt to explain . . . similarity of pattern in members of the same class, by utility or by the doctrine of final causes."[21]

Yet, although Darwin's theory of evolution is said to presuppose a randomness in the evolution of various life forms, many brilliant people have seen a teleologic propensity in his depiction of evolution, as Thomas Huxley poetically reflected:

> There is a wider teleology which is not touched by the doctrine of evolution, but is actually based on the fundamental proposition of evolution. This proposition is that the whole world, living and not living, is the result of the mutual interaction, according to definite laws, of the forces possessed by the molecules of which the primitive nebulosity of the universe was composed.[22]

What was it that resulted in living forms? We now know that it was energy, the phenome*non* that sprang forth in an insurmountable explosion, the big bang, which resulted in this universe and all of its living and nonliving constituents. It is plausible that human beings would evolve, like all other things in this universe have evolved. It has been the complex, continual transformation of energy, into innumerable forms, which finally resulted in human beings. We are but a form of energy, much the same as the entire universe. Only we are tangible, perceptible, manifestations of energy, with the inherent ability to think, to feel, and to make choices.

It is fascinating that in so many ways living things are similar. The view that there is an underlying unity of form prevalent in all living things is not a new one; in fact, as previously discussed in some detail, Aristotle and many others have suggested that this was the case. Take humans, for example. We all have a heart, lungs, brain, etc. Yet despite the orderliness and commonalties, each individual is unique; the way we walk, talk, think, and in fact look is unique for each of us. The rich diversity exhibited among living things is a reflection of energy manifesting itself in a multiplicity of ways.

Disorder or Order?

There is a topic that has been omitted from the discussion thus far and can be evaded no longer. The topic is entropy, a concept inherent in the second law of thermodynamics that essentially means there is a tendency for any (isolated) system to move toward a state of increasing disorder. On a more general basis, entropy implies that there is a certain amount of energy in any *closed* system that is unavailable for work processes. And as the amount of entropy increases the level of energy available for work purportedly decreases.

The reason entropy hasn't been mentioned up until now is that it didn't seem necessary to discuss behaviors within a closed or isolated physical system. For it is apparent that when one speaks of our universe we are referring to an incredibly open and even infinite thing. Moreover, the concept of entropy is intricately related to the idea of time, which opens up another wide area for controversy and philosophical discussion. Although these topics have not fit into the general realm of things thus far, with the introduction of chaos theory, an area that cannot be omitted, time and entropy become important concepts to at least touch upon.

Richard Schlegel had discussed the relation between thermodynamics and time, noting that "the progression of states of continually increasing entropy is the same as a past-to-future series in time."[23] So generally it has been believed that the entropic — running down disorderly tendencies of things in the universe — give us a unidirectional arrow of time. Schlegel also notes:

> The course of change in the natural world, as time goes on, is characterized by a general increase in the entropy of the various systems making up the world. Only if there were no change there would be no continual entropy increase; but in this case, we also would not have time.[24]

The recognition that many highly entropic systems do not display increasing entropy but do change nevertheless opened wide for discussion the whole notion of time's arrow. In fact, if there is change without increasing entropic tendencies, one must question if time can exist at all in any sense other than the present "now." It becomes impossible for there to be a unidirectional arrow of time based on the idea that increasing entropic change is a past-to-future time series, because then any other kind of change, especially where entropy does not increase, would logically entail a different time series. And what time series is left? Possibly a future-to-present or past series, and of course many of us would consider that absurd.

Without going further into the aspects of reversibility, or irreversibility for that matter, thus setting aside the notion of the arrow of time, we need to briefly focus on the idea of energy in relation to the new science of complexity. It seems that many dynamical systems that were thought to exhibit high amounts of entropy (unusable energy), can now be partly explained by chaos and complexity theories. That is, despite the seeming randomness and disorder there exists within certain systems an inherent ability to self-organize, although perhaps in a disorderly sort of way. The key to this relatively new area of study is the recognition of an underlying pattern in a system thought to have a high degree of entropy. We have systems that display high entropic tendencies when all the same there is increasing order. Chaotic systems are found at all levels of nature, not just on a microscopic level.[25]

The term "chaos," ever since antiquity, has often denoted a complete lack of form or randomness. So it is not surprising that in modern science the term is frequently used in much the same manner. But since the 1970s the term "chaos" has acquired several meanings, and, as scientist Edward Lorenz notes, one of the most common meanings of the term "chaos" refers to processes that may look random, but in essence are not random. Apparently many processes that were thought to be random are now said to be *"sensitively dependent on initial conditions."*[26]

The idea of sensitive dependence is nowhere better expressed than in the famous example offered by Lorenz in a lecture he presented in 1972 entitled "Predictability: Does the Flap of a Butterfly's Wings in Brazil Set off a Tornado in Texas?" In this lecture he proposed that the atmosphere is unstable and consequently it may well be that "two particular weather situations differing by as little as the immediate influence of a single butterfly will generally after sufficient time evolve into two situations differing by as much as the presence of a tornado."[27] Whether or not this is actually the case no one knows. And as Lorenz points out, the history of the expression "the butterfly effect" is perhaps a bit muddled.[28]

Whereas in the past science had focused on deterministic, entirely predictable systems, now the focus has shifted to order arising out of disorder, which entails a system being in essence predictably unpredictable. Along with the growing new science of self-organizing complexity has come the knowledge that different systems share certain commonalties. Therefore, although long-term outcomes of chaotic systems cannot be predicted, there is evidently a certain pattern, a universality, among the many different systems. So, as scientist Stuart Kauffman notes, when discussing chaotic systems, the term "universality class" is often used, referring to the fact that the same discernible patterns of self-organization exist among a number of different systems.[29]

Creative Evolution: The *Élan Vital* and Dissipative Structures

Earlier I had mentioned the Western philosophic view of vitalism in conjunction with the concepts of *ch'i* and *prana.* One of its leading proponents was French philosopher Henri Bergson, who proposed that evolution was a creative process in which the past, present, and future are somehow merged into one. Bergson used the term *élan vital* to depict the driving force which propels life in a direction toward higher levels of complexity. According to him change was directed by something but the outcomes were not predictable; thus, he rejected the Aristotelian idea of final causes. Nevertheless, his *élan vital,* which is another name for a natural vital *force* that lends direction to things, is certainly a reoccurring theme linked to the energy enigma.

The *élan vital* can be likened to the concepts of *ch'i* and *prana* since they essentially posit that there is a vital energy inherent in living things. Accordingly, a similarity does exist between Aristotle's final cause and the *élan vital;* the final cause is an inherent aspect of something which directs it to become what it eventually does. The difference is that while the *élan vital* lends direction to increasing complexity, it does not predict the outcome, while the final cause often does. However, perhaps the most startling resemblance to Bergson's ideas is that which exists between the *élan vital* and the theory of dissipative structures.

Nobel Laureate Ilya Prigogine discovered that certain chemical reactions self-organize into complex patterns, and that irreversibility is a fundamental aspect of nature.[30] This self-organizing ability cannot be explained by linear and thus determinate equations. As in chaos theory, the premise underlying dissipative structures is that order can and does arise from disorder. Whereas in classical thermodynamic systems there is a tendency to maintain equilibrium, in dissipative structures there is a tendency to move toward greater disorder (non-equilibrium) and at the same time greater complexity. This can be surmised as follows: dissipative structures appear to break down asymmetrically or unequally; however, the breakdown results in a *self-organized,* more complex thing.

Erich Jantsch notes the seemingly self-willed character of dissipative structures, pointing out that they are "self-referential," that is, dissipative structures change in a manner conducive to self-renewal.[31] Because dissipative structures are open systems, there is a continual, consistent fluctuating change in the system itself, often leading to what appears to be spontaneous new levels of order. Hence, dissipative structures autonomously self-organize, in other words, it appears that their behavior is purposive. Jantsch elaborates on the ostensibly conscious kind of behavior that dissipative structures portray:

> If consciousness is defined as the degree of autonomy a system gains in the dynamic relations with its environment, even

> the simplest autopoietic systems such as chemical dissipative structures have a primitive form of consciousness. . . . And a dissipative structure "knows" indeed what it has to import and export in order to maintain and renew itself.[32]

So we could say that things seem to *know* what they must do to actualize potentials and that they do it by transforming or changing accordingly. Perhaps then what appears to be the importing and exporting of some "thing" is actually the changing or transforming nature of *one* thing, energy. For a dissipative structure is more than just *in* a dynamic relation with its environment. *It is an intricate part of the environment, the whole.*

Just as Bergson's *élan vital* propels things toward higher levels of complexity, dissipative structures self-organize into greater levels of complexity. And, even though the final state of a dissipative structure cannot be predicted, the structure still moves toward and attains order rather than disorder. There is purposeful movement in its becoming what it does. So while both Bergson and Prigogine do away with final causation, the end results of both the *élan vital* and dissipative structures are *organized* and *meaningful* entities.

Are There "Hidden Variables"?

When we combine our knowledge of the microscopic world with that of the macrocosm, it becomes difficult not to ponder the question of whether there is some kind of hidden variable, some "that for the sake of which," underlying the becoming of it all. Indeed, many individuals have spent their lifetimes trying to uncover the mystery of this wonderful universe, an endeavor that has led to a number of hidden variable theories which try to explain the apparent unity and purposiveness exhibited on all levels throughout the cosmos.

The Implicate and Explicate Orders

David Bohm is one such scientist who postulated that there are hidden variables, which he called the implicate and explicate orders. The implicate order is said to be the foundation of everything in the universe and accounts for the integrated wholeness, while the explicate order is all that is unfolded or discernible to human beings, giving rise to our own level of existence. The implicate order is enfolded and is the fabric out of which the explicate order is evolving.

According to Bohm, the word "implicate" means to "fold inward." On the other hand, the explicate order is that which is perceptible to us. Bohm gives this example:

> In a television broadcast, the visual image is translated into a time order, which is "carried" by the radio wave. Points that are near each other in the visual image are not necessarily "near" in the order of the radio signal. Thus, the radio wave carries the visual image in an implicate order. The function of the receiver is then to *explicate* this order, i.e., to "unfold" it in the form of a new visual image.[33]

The implicate order is similar to Aristotle's idea of form and, thus, the final cause (the inner principle of change), which accounts for the tendency of things to actualize and become what they do.

Science writer Michael Talbot poetically explains Bohm's theory:

> Because the implicate order is the foundation that has given birth to everything in our universe, at the very least it also contains every subatomic particle that has been or will be; every configuration of matter, energy, life, and consciousness that is possible, from quasars to the brain of Shakespeare, from the double helix, to the forces that control the sizes and shapes of galaxies. And even this is not all it may contain.[34]

Bohm believed that a system might begin with a random change which establishes a tendency toward a series of similar changes. He held that these tendencies constitute an internally ordered process of evolution. This enfolding and unfolding of implicate and explicate orders denote rhythmical patterns which are inherent in nature.[35]

What apparently started out for Bohm as a reconceptualization of things relative to the quantum world blossomed into a new conceptualization of the universe and everything in it. He likened the entire universe to a hologram, a photographic image that may appear distorted or blank to the naked eye but nevertheless has a hidden or enfolded type of order. According to Bohm, within each part of the hologram there is an implicate order of wholeness since "each part contains information about the *whole object*."[36] In a profound way, the holographic universe *is* eternal because everything that is, was, or ever will be is in it now.

Bohm theorized that there is an infinite sea of energy filling up all of empty space and that everything in existence has come to be from this space full of energy. The infinite sea of energy is considered to be part of the multidimensional implicate order. To emphasize that everything is an undivided whole, he used the term "holomovement" to indicate something which carries the implicate order. The holomovement is "an unbroken and undivided totality" and it is ultimately *"undefinable and immeasurable."*[37]

The idea of a holographic universe reminds me of the "all is one" thesis proposed by Parmenides. If *everything* is a manifestation of energy, in an essential way, all *is* one. All is one in the sense that everything has derived from the phenome*non* of energy, and as it actualizes potentials, we can identify the unique energy forms that are manifested. But all of the energy forms are never independent entities; they are constantly changing and as such are part of a larger, indivisible whole. To view the universe as an integral whole which cannot be divided into parts in any way other than in an abstract sense poses a real challenge for science.

Consciousness, a Hidden Variable?

Consciousness has also been proposed as a hidden variable in the becoming of it all. Two modern theories, which maintain that an all-pervading kind of sentience is somehow directing the actualizing of potentials throughout the cosmos, are astronaut Edgar Mitchell's dyadic model and philosopher Mark Woodhouse's energy monism. Each of these theories suggests that consciousness is an innate aspect of the universe.

Mitchell offers a dyadic model in which, although all of nature is said to be a single reality, it is composed of physical and mental aspects or, as he puts it, existence and knowing. He set out to present his theory in two pages of his wonderful book *The Way of the Explorer*, and perhaps he met his goal in even less than that:

> Energy, we know, is the foundation of all matter; information is the foundation of knowing. Both were present at the moment of creation. It is likely that just as energy produced the

> physical structure that we recognize as waves and particles in our macroworld, the seeds of consciousness were also present to produce awareness and intentionality.[38]

Mitchell enlists the theory of the zero-point field, an idea that has been around in physics for a while. He defines the zero-point field as "the field of quantum fluctuations that exists at a temperature of absolute zero," but the terms "zero-point energy," "vacuum energy," and "an all pervasive electromagnetic field," equally apply here.[39] More important than the technical aspects of this field is Mitchell's emphasis that in it lies the potential for all existence and knowing. So that while the zero-point field is said to be "outside space and time" it is believed to be the "very ground of existence," somewhat similar to Bohm's implicate order.

On a comparable note Woodhouse, in his energy monism theory, proposes that everything is made of energy and consciousness. Monism infers a denial of any kind of duality, especially the duality of matter and mind. Yet Woodhouse notes that matter is a form of energy and then adds in the factor of consciousness, which some individuals equate with a material mind. Indeed, many individuals believe that consciousness is simply a function of the brain, which is a topic for the next chapter. However, in energy monism, clearly consciousness is not simply a function of mind, but rather it is a "transpersonal field that subsumes its various contents and capacities."[40]

Woodhouse asserts that neither consciousness nor energy can be reduced to the other, but that each is somehow reflected in the other. He unites energy and consciousness into one thing by asking us to think of energy as the outside of consciousness and, conversely, consciousness as the inside of energy.[41] In energy monism the physical and nonphysical are one, rather than two disparate realities. In his view, "consciousness is the seat of meaningful or intelligent patterns of information which are expressed energetically over space and time."[42] This implies that energy is used to somehow transport or move consciousness. But he also stresses that "energy is the fundamental reality," whereas matter is only a derivative state.[43]

Essentially, it appears that the dyadic model and energy monism entail the existence of two fundamental things: energy *and* consciousness. Here we could appeal to Ockham's razor — the principle that says that we should always try to explain things using the fewest concepts because the simplest explanation tends to be the right one. So much simpler, it seems, to believe that all is essentially one, and that consciousness, in fact all of our mental abilities, is but another meaningful manifestation of the phenomenon, energy. In the next chapter we will explore this idea further.

Morphogenic Fields

Around the time that all the extraordinary things were happening in physics in the early part of the twentieth century, it came to be known that living things are also comprised of and organized by fields.[44] These fields, termed morphogenic, are proposed by scientist Rupert Sheldrake to be the cause of the final developed form of a living thing. Not surprisingly, Sheldrake likens morphogenic fields to Aristotle's formal and final causes:

> These organizing fields . . . play the role of formal and final causes. They give the developing organism the end or the goal toward which it is developing, and the final form it is going to develop towards. . . . It has a kind of purposive quality.[45]

Interestingly, he emphasizes that these morphogenic fields are not energetic.

Sheldrake challenges the traditional idea that nature is somehow governed by mathematical equations which lie outside the realm of being. In answer to the question of how living things become what they are, he uses the idea of morphic resonance, "the influence of like upon like."[46] Morphic resonance is said to be different from that of energetic resonance in that it does not involve any energy exchange or transmission. The morphogenic field is said to contain a memory of the past which links the past with the present and future becoming of a particular thing; time and space collapse in a kind of all-pervading memory field. Thus, the cause of becoming is within the field itself, rather than outside it.

Concerning the idea of morphogenic fields, Briggs and Peat notably linked Sheldrake's fields with other ideas such as Bergson's *élan vital,* Driesch's *entelechy,* and auric fields. Morphogenic fields share common elements with Aristotle's *entelecheia* and Leibniz's *entelechy* as well. The idea is also moderately similar to psychologist Carl Jung's idea of the collective unconscious, a part of the mind that is said to be universal having "contents and modes of behavior that are more or less the same everywhere and in all individuals."[47]

The Earth as Gaia

There are those who believe that the earth is a self-organizing entity, a kind of superorganism that intimately coexists with all living things.[48] Named Gaia, after the Greek earth goddess, this theory basically proposes that all is one on a planetary scale. The idea of the earth as Gaia, first proposed by atmospheric chemist James Lovelock and microbiologist Lynn Margulis, is supported by the fact that there are many regulatory mechanisms which seem to maintain a balance conducive to the continuation of life on earth. For example, the earth's temperature seems to be self-regulated and in a way that has been entirely favorable and conducive to life formation.

Lovelock pointed out that our planet has many homeostatic mechanisms, too many for it to be just a vast rock in the universe. However, his proposal of the earth being akin to a living organism has met with mixed responses. For instance, John Polkinghorne, a physicist and Anglican priest, believes that "the Earth looks neither like a machine nor like a living thing, but like some form of self-regulating entity lying in between."[49] We could say that the earth is simply another meaningful manifestation of energy; another integral part of the wholeness of the universe coordinating its efforts with everything else in the becoming of it all.

Summary

A number of ideas have come to light. First, energy is a thing — a phenome*non* — that has given rise to everything that exists. On that account, there is an overarching idea that all is essentially one; at the very least it is one in the sense that everything has arisen from one universal phenomenon, energy. Such being the case, everything is connected in a way that cannot ever be reduced to parts. Change, then, is an inherent characteristic of the *one* phenomenon, energy; it comes about through communal process. And last, it seems that there is an undeniable, teleological quality to the phenome*non* of energy.

If everything is a manifestation of one thing, energy, it would appear that the process of change arises from within rather than from without and is therefore intrinsic to the phenomenon, energy. Now it remains for us to speculate; if everything we see, and even what we cannot see, is a manifestation of energy, can we then doubt the purposiveness of this phenomenon? This question leads us to the perhaps most significant point: self-organization.

In chaos theory we have a subtle indication that there may be formal and, thus, final causes in nature. Chaos theory proposes that order can arise out of disorder. Although unpredictability is found to be even more evident in nonlinear systems, amid this seemingly random behavior one still finds pattern and self-organization. For instance, journalist Malcolm Browne reported research findings indicating that when granular objects such as sand or even brass balls are set into motion they tend to self-organize into complex patterns.[50] This seems to imply that energy—which is at the root of chaotic systems (since chaotic systems are manifestations of energy) and their inherently changing patterns — is purposeful. In essence we could say that the organized patterns arising out of chaos are indicative of some kind of inner principle of change.

With an examination of the becoming of it all the idea of energy having a distinct self-organizing ability seems even more apparent. Jantsch discussed in great depth the obvious self-organizing ability of the universe along with everything in it, and although his focus was not on the phenomenon of energy, he made an interesting comment about dissipative structures:

> It may be interesting to ask whether a dissipative structure is to be understood as a material structure organizing the flow of energy, or as an energetic structure organizing the flow of matter. At this level of self-organization, both descriptions are equally valid. They constitute two sides of a complementarity. At higher levels of self-organization, however, to an increasing extent a description will suggest itself which views energy systems manifesting themselves in the organization of material processes and structures.[51]

Although I may be taking Jantsch's comment a bit out of context, it does help to elucidate an idea that has emerged concerning the phenome*non* of energy. If matter and energy are interchangeable, then any material structure, if it is organized, is organized by the flowing, transforming of energy. In cases like dissipative structures there is an apparent self-organization of the material aspects of the energy form. This self-organizing ability seems to be everywhere evident.

Along with the apparent self-organization there necessarily comes a distinct notion that energy has a *self-determining* ability. Prigogine and Stengers in *Order Out of Chaos* note: "Our universe has a pluralistic, complex character. Structures may disappear, but they may also appear."[52] So, we could ask *why* do material structures disappear and appear? It seems they often do so *at will.* Apparently various manifestations of energy are self-organizing and consequently determine where and what they become.

It seems that our universe and every aspect of it is fairly strong evidence that the forms of energy work together harmoniously toward some mutual goal, which I have termed "the becoming of it all." If everything in nature is viewed as a manifestation of the one phenomenon, energy, then we truly have a self-organizing *whole.* And, just as a tide flows unceasingly in and out, energy transforms, flowing within and without, backward, forward, in every direction.

Recently, a team of international astronomers reported findings that add credence to the belief that there may be pattern and organization in the heavens. Specifically, a group of astronomers indicated that superclusters or what are called giant galaxies appear to be arranged in a pattern that resembles a three-dimensional chessboard.[53] If this turns out to be so, people might no longer so easily assume that chance brought about the heavens. As George Johnson notes: "Maybe, back in the beginning, the dice were loaded. As the universe was unfolding, some unknown ordering principle may have been at work."[54] What's more, researchers strongly indicate that the chances of our coming into existence — meaning our world with all of its forms of life — were slim. If events during the big bang or bang(s) and thereafter had not occurred precisely as they did, we would not be here.

Indeed, the primordial energy of the big bang (or perhaps bangs) and the subsequent actualizing of potentials do seem to indicate that some inner principle of change existed before the matter and then within the matter, once actualized. It's almost as if energy has a mind of its own, a sentience about it that directs itself toward further complexity, an idea first proposed by scientist and philosopher Pierre Teilhard de Chardin. It is becoming.

Take a tree, for example. It came to be, but it is not simply being; it is becoming — all the time. And why is it that an evergreen stays green in winter while a deciduous tree loses its leaves? No, I don't mean the reason — the purported causal explanation for this to be so. I mean *why?* I'll bet there is a reason, a purpose underlying this phenomenon — the tree — but it isn't something that we humans can discern. It isn't simply that there has to be a purpose *for us,* since we do not, after all is said and done, exist in a vacuum. But it seems it must have a purpose. We are only a part of the indivisible wholeness. Everything in the universe is becoming; however, it is all becoming together. It is all energy, and, as such, it is all essentially one.

One can't find fault with those such as Bohm, Sheldrake, Mitchell, or Woodhouse for hypothesizing that there are hidden variables underlying our existence. It does seem so. There is a purposiveness in all things, even the smallest and inanimate things. The ground of existence implied by Bohm's implicate order and the idea of the zero-point field, are they possibly, like morphogenic fields, a depiction of Aristotle's final cause, scientific terms employed to try to quantify the tendency for things to become what they eventually do?

In Aristotelian terms, every actuality (*energeia*) is actualizing its potential in a self-organizing sort of way. The evidence we see all about us leaves it extremely difficult to believe that the becoming of it all can be attributed to chance, yet as Sheldrake notes, many have claimed that this is so:

> There are a variety of theories of creativity. The one that comes cheapest is the materialist theory: It all happens by chance. For as soon as you've said that, there's no more to be said. It simply says that we can't understand it.... It's chance. It's random.[55]

Sheldrake doesn't see any particular reason for believing this, and, quite frankly, neither do I.

The next two chapters relate specifically to human beings. Theories concerning the nature of sentience, health, and healing will be examined in a further effort to shed more light upon the enigma of energy. It is best to keep in mind that these areas, in essence, focus on the health and well-being of humankind. And as we all know, health is itself an enigma.[56] We don't really know how or why living things are so well put together. Nor do we know why they often heal without any kind of intervention. But with a new set of glasses on, which enables us to see that everything is a manifestation of one thing, energy, it really shouldn't surprise us that this is so.

Chapter 4

Human Beings: More Than Just Matter

> To develop great men, such as Aristotle, Plato, Shakespeare, Milton, Washington, Lincoln, and then by death to quench them in utter oblivion, would be unworthy of Omnipotence. To my mind, it is simply an impossible conclusion. Man's soul *must* be immortal. — WILLIAM KEEN[1]

We humans are truly an exquisite species. We love, we hate, we learn, we hope, we dream. We are rational, thinking beings, though we can be irrational as well. That is to say, we think, and we make choices, some of which make sense to most, and some of which make sense to few. The mind is by far the most enigmatic of our human qualities, particularly its conscious aspects such as awareness, emotions, and intellect, but also the *subliminal,* subconscious things, such as dreams and intuition. In fact, many great men and women have claimed that it was their intuition or dreams that guided them.

It is suggested that by focusing on the idea of energy we may gain further insight into the dilemmas encountered when we attempt to make sense of our conscious and subliminal abilities — the ability to think, to feel, to sometimes know without knowing, and quite simply, to be more than just matter.

The Mind-Body Problem

The mind-body problem, plainly stated, is the question of whether the mind and the body are one or two distinct entities. There are many individuals, myself included, who question the very premise that the human mind exists within the confines of the physical body. And, while there is a general understanding of the functioning of the human body, conscious and unconscious processes are still not well understood. In fact, little is known them.

Interwoven with the question of whether the mind and body are two distinct entities is the question of immortality. There is a pervading idea that consciousness, or what many people refer to as the soul or human spirit, may not be limited to the confines of a body, and therefore it, the soul, might survive the demise of the physical body that always occurs after a period of time. The idea of immortality follows almost logically if one assumes that the mind and body are separate entities. However, in the mind-body arguments, it doesn't always follow that if the mind and body are one, immortality needn't be or can't be real.

The mind-body problem can be traced in Western thought as far back as the ancient Greeks, and it especially comes to the forefront with Plato and Aristotle. Most of the ancients used the word "soul" to describe not only mental capacities but also things such as nutrition and growth. In Plato's works, particularly his dialogues, there is an inescapable distinction between the body and mind (soul).[2] However, he did believe that there must be harmony between the two: "... that we should not move the body without the soul or the soul without the body, and thus they will aid one another, and be healthy and well balanced."[3] Plato viewed the soul as an enduring entity; it existed before and after the body.

On the other hand, as far as Aristotle was concerned, there was no mind-body dichotomy, for the two were parts of one whole. Recall Aristotle's antitheses of actuality-potentiality and matter-form. The flesh and bones, the body of the human being, comprises the potentiality, and the *soul* of the human being is the *first entelecheia* (an *energeia*). The soul is both the formal and final cause of being human; it is the guiding influence that gives matter its potential to actualize further. A living thing would simply not be if it were not for the soul, that aspect which gives a living thing its nature. As such, when a living thing dies, so too does its soul.

It does logically follow that if the soul (as form) is in the physical body, then when the body dies, the soul dies as well. But there is more to be said about Aristotle's soul. The soul is in essence the powers or capacities of living things to actualize or to function.[4] In his work *On the Soul* Aristotle speaks in depth about the powers of living things, including nutrition, perception, appetition, change in place or movement, and thought. Of all these capacities that Aristotle believed belonged to the soul, he singled out one, that of *thought,* to be *immortal.* The rest of the capacities are physical entities and as such must die with the physical body.

Because the power of thought, on the one hand, is of an active nature and, on the other, can be of a passive nature, Aristotle believed that it must necessarily be different. The active nature of thought must be a part of the potential in each of us to actualize our innate abilities. For that reason, thought must be eternal; it must be an actuality before it can become a part of the soul in any living thing. To borrow a quote of Aristotle's reprinted by Jonathan Barnes: "Hence it remains that thought alone comes in from outside, and that it alone is divine; for corporeal actuality has no connection at all with the actuality of thought."[5] So Aristotle's conception of thought as an enduring actuality, an *energeia,* is congruent with the modern idea of an eternal soul.

Descartes later expanded the Platonic view that the body and mind are separate entities. In Descartes's view, the mind is nonphysical; one of his principles of philosophy — "I think, therefore I am" — captures his rationale for believing this to be true. In other words, we can doubt the existence of all things in this world, with the exception of ourselves, since we "perceive that we think."[6] Consequently, he drew the conclusion that the soul continues to exist without the physical body: "that this 'I' [that is to say, my soul, by virtue of which I am what I am], is entirely [and truly] distinct from my body and that it can [be or] exist without it."[7]

Although Descartes, like Plato, argued that the mind is a distinct, separate entity, like Plato, he acknowledged that the mind works in conjunction with the corporeal or physical body: "Nature . . . teaches me by these feelings of pain, hunger, thirst, and so on that I am not only residing in my body, as a pilot in his ship. . . . I am intimately connected with it, and that [the mixture is] so blended, as it were, that [something like] a single whole is produced."[8]

Leibniz, similar to Aristotle in his thinking, argued against the Cartesian mind-body dichotomy. Leibniz believed that the mind is inseparable from the body; his idea was contingent upon his concept of entelechy: "It appears that each living body has a dominant entelechy, which in an animal is the soul."[9] Thus, the entelechy of a living thing is its soul. According to Leibniz, each portion of matter, each monad, contained within it a multiplicity of parts which in turn can be infinitely subdivided into other parts, all of which, for Leibniz, constituted a whole: "And though the earth and the air which are between the plants of the garden, or the water which is between the fish of the pond, be neither plant nor fish; yet they also contain plants and fishes, but mostly so minute as to be imperceptible to us."[10] So within each monad there is, in an ideal sense, an expression of the whole universe.

Leibniz also held that there was no such thing as a void and concurrently perceived that there is "nothing dead in the universe, no chaos, no confusion save in appearance." His reflections show an uncanny insight on his part: "All bodies are in a perpetual flux like rivers, and parts are entering into them and passing out of them continually."[11] The body and soul are inseparable to the extent that there is no "absolute birth" or "complete death." The soul is immortal.

For Bergson, consciousness was an enduring actuality, retaining the past as it moves into the future:

> Continuity of change, preservation of the past in the present, real duration — the living being seems, then, to share these attributes with consciousness. Can we go further and say that life, like conscious activity, is invention, is unceasing creation?[12]

He used the metaphor of a "central telephonic exchange" to depict the brain. The entire nervous system is simply a conductor. The matter — the brain — works with the mind; the mind uses the brain in order to actualize potentials. The idea of the brain as a mere conductor could lead us to assume that the mind needs the body in order to actualize certain potentials, which could consequently lead us to suppose that the body and mind are indeed one. Yet, Bergson made a vital distinction: "Mental life cannot be an effect of bodily life, that it looks much more as if the body were simply made use of by the mind, and that we have, therefore, no reason to suppose the body and the mind united inseparably to one another."[13]

Bergson also believed the mind continues to exist after the physical body dies:

> If the work of the brain corresponded to the totality of the consciousness, if there were equivalence between the cerebral and the mental, consciousness might be bound up with the brain and death might be the end of all. . . . But if, . . . the mental life overflows the cerebral life, if the brain does but translate into movements a small part of what takes place in consciousness, then survival becomes so probable that the onus of proof falls on him who denies it rather than on him who affirms it.[14]

Teilhard de Chardin believed that life is a sign of a universal process of ever-growing complexity and increasing consciousness. According to him, consciousness existed at every level from the atom to the most complex, that is, human beings. Seeing a fundamental harmony between what he referred to as spiritual energy and matter he noted, "*somehow or other,* there must be a single energy operating in the world" and that in essence this "energy is psychic in nature."[15] This psychic energy is not a new form of energy; rather, it is a different way of viewing energy.

The belief that all energy is psychic lays the foundation for what Teilhard called hyperphysics, a theory in which he postulated two fundamental components of psychic energy: *tangential energy* (the physical activity of energy) and *radial energy* (the inherent ability of energy to move toward greater complexity).[16] He rationalized that if energy is psychic in nature, it would account for the development of increasing complexity and consciousness in the universe. For that reason he strongly avowed: "Under penalty of being less evolved than the ends brought about by its own action *universal energy must be a thinking energy.*"[17] Consequently, he believed that the unity of all things and their ongoing transformation can only lead to a transcendence of human consciousness that will rise beyond material being. We are gradually transcending the material confines of our universe.

The Mind Is in the Matter: The World of Neurons

A good deal of the scientific literature related to sentient abilities deals with the microcosmic world of neurons, which is a form of matter often said to be what enables humans to think and feel, to move and act, to be, in the most basic sense, human. In this view, we are simply matter. Science writer and astronomer David Darling, who argues against this view, notes:

> What, then, can we hope for after death? Nothing — absolutely nothing — if we believe what many scientists say. All life, they argue, can be understood in terms of chemical reactions. Every event, all the wonders of nature, can be explained by the accidental bumping and jiggling of particles. The brain *is* the mind. Why trouble to speculate further about an immaterial soul or afterlife?[18]

The belief that mental processes arise from the brain falls within the realm of materialism and was first strongly advocated by psychologist Gilbert Ryle. In a rather strict sense, materialism completely does away with the mind-body problem because there is only a body.

Included in the group of individuals who believe sentience is a function of matter is philosopher Owen Flanagan. He notes: "The most plausible hypothesis is that the brain is a Darwin machine, a massively well-connected system of parallel processors interacting with each other from above and below and every which way besides."[19] Flanagan is quick to note what he believes to be the main disadvantage of Descartes's mind-body dichotomy: "We have no evidence whatsoever that there are any nonphysical things."[20] When I read statements like this I can't help wondering about all the elusive wave-particles out there, not to mention what physicists refer to as the *virtual* ones. And isn't it very hard to characterize those nonlocal connections between split photons as physical?

One of today's leading proponents of the "mind is in the matter" view is scientist Daniel Dennett. He believes that the problem of consciousness has already been explained.[21] The mind is the brain, period. Dennett has been lending assistance to the development of a robot named Cog. It is hoped that one day Cog will be able to simulate human beings in more ways than artificial intelligence has been able to do thus far. For example, Cog is being built to have something like skin so that it can be programmed to react to outside factors, such as heat. Just as humans withdraw from heat in order to prevent trauma, Cog will be programmed to react in the same manner. Dennett believes that forms of artificial intelligence such as Cog help to explain consciousness.

But then there are the New Mysterians, a group of scientists and philosophers who believe that consciousness is not just the material of the brain; it has a mysterious aspect to it as well. One of the leading proponents of this group is David Chalmers, who in his search for a fundamental theory of consciousness tries to do away with the materialist view, noting that there is an undeniable dualism; however, he emphasizes that he is not referring to Cartesian dualism. He maintains that "consciousness is a *feature* of the world over and above the physical features of the world. That is not to say that it is a separate 'substance.' "[22] According to Chalmers "consciousness supervenes naturally (although not logically) on the physical."[23]

Opposing Dennett's view, Chalmers believes that Cog and other forms of artificial intelligence don't and won't explain consciousness; on the contrary, the mystery will be deepened. For no matter what Cog does, Cog cannot *feel.* Cog might one day be programmed to recoil from heat and other threats to its "skin," but it will never be able to feel or, thus, express heat or pain.[24] Noting a coherence between awareness and consciousness he hypothesizes that this apparent connection is a fundamental law of nature. Thus, Chalmers assumes that functional processes (awareness) seem to be necessary correlates for consciousness: "So far, all we know is that consciousness arises from the physical somehow, but we do not know in virtue of what physical properties it so arises."[25]

Another prominent figure in the study of consciousness is Karl Pribram.[26] He likens the human brain to a hologram, noting that memories are nonlocal; they are located throughout the brain rather than in discrete local parts of the brain structure. Nevertheless, he also appears to advocate a neurophysiological solution to sentience. Yet, the holographic view expounded by him and Bohm, in which time and space are collapsed, provides a possible explanation for events such as precognition and telepathy which are remarkably similar to the nonlocal phenomena known to occur in the microworld. It is somehow hard to reconcile processes *within* the brain with these para-phenomena.*

---

*"Para-phenomena" is my word for things that are currently beyond our empirical comprehension, i.e., split photons that know what the other is doing, nonlocal healing, etc.). The word "parapsychological" has cropped up in recent literature in lieu of the word "paranormal." "Para" means beyond; coupled with normal, it means beyond normal. But what *is* normal? More and more it seems that many things we thought were abnormal

Roger Penrose, a well-known mathematical physicist, sums up a majority of viewpoints currently being expressed in the sciences concerned with the phenomenon of awareness. He advocates that "awareness cannot be explained by physical, computational, or any other scientific terms" and gives the illustrious example of Deep Thought, a very powerful computer that had lost a game of chess to a human being.[27] It is generally assumed that the reason the computer lost was because, while the computer can compute the moves necessary to play a game of chess, it is unable to *understand* the game. But recently I heard a news broadcast about the world champion chess player who finally did lose a game to a computer. So it would seem that humans have finally succeeded in *programming* a computer to win at the game of chess. Penrose contends that a possible answer to the mystery of consciousness lies in a new physics, that is, in a form of noncomputational, although still somehow physical law.[28]

Beyond the Brain or Beyond the Body?

In the Hippocratic works (fifth and fourth centuries B.C.) there was a reference to the idea that consciousness permeates the entire body; that is, as renowned physician Larry Dossey says, there was an idea of a "body-wide mind." Recent investigations in neuroscience lend support to the idea that consciousness is dispersed throughout the body. Studies have found that endorphins (chemicals that supposedly affect our mental moods) previously believed to be found only within the brain, exist in many other parts of the body such as the gastrointestinal system.[29]

Dossey reports patients who demonstrate a kind of distant knowing. He gives a particularly striking example of a female patient who had come to him because of an enlarged ovary. Another physician had suggested that she have the ovary removed since it might be cancerous. But somehow she *knew* that it wasn't. She *knew* that there were "three little white spots on it, and they are not malignant."[30] A sonogram later confirmed that she was right.

---

are normal. Also, nonlocal microcosmic events *used* to be considered "paranormal." The word "para-phenomena" seems appropriate for all these events.

Numerous reports indicate that the mind communicates with distant parts of the body. Often these reports are linked with ideas of healing which will be discussed in the next chapter; however, there is one form of distant knowing that coincidentally ties in not only with the topic of sentience, but also with that of the eternal soul. The form of distant knowing I am referring to is the near-death experience (NDE), or the experience of coming back from the dead. As far as I can tell, near-death experiences fall on the borderline between mysticism and theology because there is no scientific way to explain them.

In 1975, Raymond Moody brought to public awareness the phenomenon of the near-death experience with his book *Life after Life*. It has since been estimated that of every hundred persons who have been declared clinically dead and thereafter come back to life, approximately forty percent report a NDE. Many have proposed explanations for the NDE; however, they all agree that none of the explanations offered thus far is adequate. Take, for example, cosmologist Carl Sagan's attempt to explain the NDE by likening it to the natural birth process.[31]

There are many commonalties reported by people who have had NDEs. Almost without exception, people who have had NDEs report being in a dark tunnel; there is a bright light at the end of the tunnel which appears to be some form of a being; and they feel peaceful and have an out-of-body sensation. Sagan had noted similarities to the birth experience: the fetus has to travel down the birth canal (a dark tunnel) and into the bright light of the world and the hands of the practitioner delivering the baby (the bright light that resembles a "being"). But then it was found that people delivered by Cesarean section, who never traveled down the birth canal, as well as those delivered by natural childbirth, provide similar accounts of their NDEs.

Psychologist Kenneth Ring offers an explanation based on holography theory. In fact, he draws a parallel between holographic reality and other para-phenomena such as telepathy, precognition, and synchronicities. First, Ring proposes that out-of-body experiences are real events, leading to the explanation that either consciousness detaches itself or a replica of the body separates from the person's physical body. Second, he speculates that the insights afforded during the tunnel and light experiences emerge from a holographic reality where "time and space are collapsed and where, therefore, causality can have no meaning."[32] Thus, people who have had a NDE, because they are no longer confined to the physical body, are able to view a different order of reality that encompasses things they were unable to see or perceive before.[33]

It seems that many of these people who come back to life have distant knowing abilities during their NDEs. They are able to report things they saw while they were, at the very least, clinically dead. But that's not all; they report having seen things *that they couldn't possibly have physically seen with their eyes.* One particular instance noted by Darling comes to mind, where the person who had come back reported having seen a shoe outside on an "inaccessible ledge" while she was floating out of her body; indeed, the shoe was out there on the ledge! These confirmed reports of distant knowing appear to outweigh several other theories which try to account for NDEs as hallucinations perhaps due to medications or lack of oxygen.

Another interesting aspect of NDEs is that most people report having had an enormously increased awareness of everything around them. Darling calls it an expansion of consciousness where there is "a feeling of total loss of self and of timelessness, all-at-oneness."[34] He likens the experience with that of people who meditate, for they too express these feelings. He notes: "From those who have skirted death comes this extraordinary new evidence suggesting that cognition may actually broaden and become more profound at exactly the same time the brain stops working. How is that possible?"[35]

Ring notes that NDEs seem to activate an inner potential, like a seed from which a flower blooms, and speculates that perhaps those who experience NDEs somehow needed them to provide a catalyst to their own development. Nurse educator John Phillips discusses the nature of dying and notes that "the growing number of reports of people undergoing near-death experiences signifies that the dying process plays a role in helping people to actualize their potentials."[36] It really does seem as if there is a "that for the sake of which" underlying these events since many individuals who have had a NDE do go on to lead more spiritual and fulfilling lives.

Moody revealed a number of parallels between the stages of the near-death experience and various notable writings throughout history. He points out extraordinary similarities between passages in the

Bible, Plato's works, the Tibetan Book of the Dead, Swedenborg's writings, and reports of NDEs. He reflects:

> How is it . . . that the wisdom of the Tibetan sages, the theology and visions of Paul, the strange insights and myths of Plato, and the spiritual revelations of Swedenborg all agree so well, both among themselves and with the narratives of contemporary individuals who have come as close as anyone alive to the state of death?[37]

Coincidental these similarities may be, but they are *meaningful* just the same.

Near-death experiences and other reported forms of distant knowing support the idea that sentience is nonlocal, not fixed in either space or time. Dossey reflects upon the potential implications of a nonlocal mind:

> If minds are nonlocal — if they are not bounded in space and time — this means that they are not fundamentally separated from each other. And if not separate, there cannot be, then, five billion single minds inhabiting the earth today, but *only one Mind.*[38]

Physicist Erwin Schrodinger also came to this conclusion in a roundabout way. Many individuals have alluded to a universal, eternal consciousness.

Accounts of NDE are similar to the mystical experience that many individuals have referred to throughout the ages. The mystical experience is comprised of feelings of all-at-oneness, euphoria, and a supernatural sense of the divine. Maslow spent a lifetime studying happy, self-actualizing individuals and noted that they often had what he called "peak experiences," experiences similar to the well-documented mystical experience. It seems that mystics, as well as people who meditate, the dying, creative and gifted, and, on occasion, happy individuals, often experience this strong feeling that all is essentially one.[39] It is a transcendent feeling that most of us do not ever experience while living because, as Henry Margenau notes, "It is our human lot to look at the four-dimension world through a slit-like opening. . . . Whenever the slit opens, and for some the slit only opens at the time of death, you see more than a segmented three-dimensional world."[40]

Para-Phenomena in the Realm of Sentience

In the realm of sentience there is a large body of evidence indicating that various para-phenomena frequently occur. Psychologist Joseph Rhine coined the term extrasensory perception (ESP) to encompass various abilities to perceive beyond the senses. Collectively, these abilities — telepathy (mind-to-mind communication), clairvoyance (knowing things that no else knows), precognition (knowledge of future events), and psychokinesis (mind moving matter) — are referred to as *psi*.[41] Rhine alluded to the possibility of a nonphysical process when he said:

> The very first characteristic of ESP (and psi in general) I shall mention is a most amazing one.... I refer to the fact that psi is not in any known shape or form a recognizable physical process. Physicists have attempted to offer explanations.... One finds field theories of psi, electromagnetic wave theories, and before these were current in physics still other notions were suggested on such lines as unknown cosmic forces.[42]

It has since been substantiated that people who have ESP do not use *any known form* of sensory perception.[43]

Psi lends further credence to the belief that conscious and subliminal abilities are not confined solely to the physical body. Psychokinesis is particularly relevant here because it is believed that the mind acts directly on matter. An example of psychokinesis that has been studied extensively is the phenomenon of metal bending. Often called the Geller effect, metal bending came to be widely known in the 1970s because of a television broadcast in England showing a male named Uri Geller performing metal bending and other parapsychological activities.

John Taylor, author of *Superminds,* notes, "The most important physical feature involved in bending is the transference of energy from the subject to the object. For only thus can work be done on the object to modify its shape, and work requires energy."[44] Energy is discussed in terms of usage and expenditure, but there is no empirical evidence to indicate that energy is being transmitted in cases of psychokinesis.

In the case of metal bending, the bending may occur while a person strokes the metal, in which case many individuals believe that muscular energy is used; however, the muscular work involved does not appear to be enough to cause the metal to bend. And then there are numerous, well-documented instances where metal bending has occurred *without any touching involved*. In addition, there have been numerous well-documented cases of metal bending taking place over distances.

In all instances of psi, time and space seem to collapse and have no meaning — just as in nonlocal events in the quantum world. How is it that mothers frequently know when their children are in trouble, despite being perhaps hundreds or even thousands of miles away? Why is it that we can dream of seeing an old friend we haven't seen in years, only to awaken the next day to its reality?

Meaningful Coincidences

A fascinating para-phenomenon is the frequent occurrence of meaningful coincidences. During psychotherapy practice psychologist Carl Jung noted that his patient's dreams or experiences often reflected some underlying order and meaning.[45] Called "synchronicities" by Jung because they appear to have some kind of connecting principle, meaningful coincidences are, like other nonlocal events, acausal in character:

> Synchronicity designates the parallelism of time and meaning between psychic and psychophysical events, which scientific knowledge so far has been unable to reduce to a common principle. The term explains nothing, it simply formulates the occurrence of meaningful coincidences which, in themselves, are chance happenings, but are so improbable that we must assume them to be based on some kind of principle. . . . No reciprocal causal connection can be shown . . . between parallel events, which is just what gives them their chance character. The only recognizable and demonstrable link between them is a common meaning.[46]

Jung also came to believe that sentience is not confined to the body nor to any particular position in space or time.

I have heard many stories of synchronicities over the years, and

I have seen meaningful coincidences occur when the individuals involved were not even aware of them. For example, I had a patient in the cardiac intensive care unit who had suffered a serious myocardial infarction (heart attack). He had been doing fairly well but suddenly took a turn for the worse. His wife had spent the last two days at the hospital, but had left about an hour before, seeing that he was stable. Moments after he went into pulmonary edema, she reappeared on the unit.

She told me that when she was about three blocks from home, a song, their wedding song, had come on her car radio. It was unusual, she said, to hear this old song on the radio, and she got a funny feeling that she should come back to the hospital. This was back in the days (a decade or so ago) when family members were allowed in the ICU only during brief visiting times, but after a bit of finagling with the doctors, she was allowed to be with her husband. He was still alert, but in obvious distress. She sat by his side and held his hand, talking to him through it all. He died about thirty minutes later.

Synchronicities are well-documented phenomena. On a personal level my family and I have experienced some very meaningful coincidences since the death of my father, so meaningful, in fact, that I have embarked on an effort to study synchronicities related to the loss of a loved one. I am convinced that these occurrences are also commonplace, and while working with a bereavement group I have heard many poignant accounts that seem to validate my intuition. Sometime in the near future I hope to document them. In the meantime, I will share a few of my experiences.

For years the number 122 haunted my mother. She would see it everywhere and would even wake up at precisely 1:22 in the morning. The haunting ubiquity of this number was often accompanied by a feeling of foreboding. In other words, she knew that whatever this number meant, it was not good. My mother also has had an uncanny way of knowing when something wasn't quite right with one of her children. She always seems to call me or one of my siblings whenever a major crisis is occurring. It seems that many mothers have this special sense in relation to their offspring.

The number 122 turned out to be quite significant for my family, although none of us, except my mother, had really paid any attention to it before. My father died, or at least his physical body died, on

January 22, 1995. After his death I experienced many things that I really wanted to believe were divine signs that he was still, in a sense, with us. For example, I would be thinking of him and suddenly I would see the number 122 or a song he used to sing would come on the radio. I alternated between belief and disbelief, for although I wanted desperately to believe, I've never been a particularly devout Christian.

One day about a month after my father left us, I was on my way home from an exceptionally absorbing evening at New York University. This was always a difficult trip home for me, because I was accustomed to stopping off at my parents' home to sit and talk with my father about all the philosophical things that had come to mind while at school. I still hadn't been able to reroute my way home, despite the fact that the house had been sold months before, and now Dad was no longer with us. Invariably I would find myself in tears while driving home.

On this particular evening when I was feeling a bit more desolate than usual, something incredible happened. I had stopped for a red light, and there was a white truck in front of me. In bold black letters on that white truck were the words, "Psalm 122: I was glad when they said unto me, let us go into the house of the Lord." I would probably never have read the words, had I not noted the number 122. And the words were truly significant because my father had died of cancer, and up until the last few days he had fought valiantly to live.

Jung would have called this a synchronicity or "a meaningful coincidence."[47] However, I think it was more than that. I might call the fact that one of my father's brothers also died on January 22, although in a different year, a synchronicity. I would even call it a synchronicity that when I went to visit the gravesite of my grandparents (my father's parents), having not been there since the age of seven or eight, I was directed to plot number 122.

In retrospect I wonder how, by chance, my brother managed to buy his first home with the address 220 First Avenue, a month before Dad died. And the fact that he died six hours before his birthday could also be called another "coincidence."[48] Yet these are without question *meaningful* events. On the other hand, for me the Psalm points to more than a meaningful coincidence. It bears a distinct meaning that reached out to me. For me, this kind of meaningful occurrence leads to religion and to the idea of everlasting life. Perhaps I could discount it all as just a coincidence, but since then I have had many such experiences.

Para-phenomena such as psi and synchronicities cannot be explained by the classical notion of energy. There is simply no basis for believing that energy is being transmitted from one thing to another. Jung commented on the paradoxical nature of such events:

> There are events which are related to one another experimentally, and in this case *meaningfully,* without there being any possibility of proving that this relation is a causal one, since the "transmission" exhibits none of the known properties of energy. There is therefore good reason to doubt whether it is a question of transmission at all.[49]

Thus, Jung also came to the conclusion that these events cannot be explained as a "phenomenon of energy."

Today there seems to be an increased interest in para-phenomena and in such events as synchronicities. There are scientists in almost every discipline who are trying to account for action-at-a-distance phenomena. Physicist David Peat brings to light some extraordinary consistencies concerning synchronicities and the new physics theories, for example, when he discusses physicist Wolfgang Pauli's exclusion principle: "the discovery of an abstract pattern that lies hidden beneath the surface of atomic matter and determines its behavior in a noncausal way."[50] Pauli's exclusion principle is a mathematical construct without which it is said that quantum theory could make no sense.[51] It is well known that Jung collaborated with Pauli on his work, and it is obvious that they were of like minds: both believed that there were acausal factors in nature.

It is believed that many of the newer scientific theories, such as Bohr's complementarity, Heisenberg's uncertainty principle, and nonlocality, offer possible avenues for further elucidation of various para-phenomena. Some, such as Taylor, believe they are a field phenomena. Others discuss Bohm's theory of the implicate and explicate orders in conjunction with nonlocal events with the hopes that one day his theory will be able to account for these mysterious phenomena.

Body, Mind, Energy, and Consciousness

John White and Stanley Krippner have identified a multitude of references throughout history to a mysterious life energy, a universal phenomenon which they call the "X energy."[52] Beginning with the concepts of *prana* (Hindu Yogis, 3000 B.C.), *ch'i* (Chinese, 3000 B.C.), and *ki* (Japanese, 3000 B.C.), they go on to list many others, such Bergson's *élan vital* and Hippocrates' *vis medicatrix naturae* (healing force of nature), as references to this mysterious life energy. The idea of a healing *force of nature* forms part of the basis for the next chapter and will be discussed in some detail.

These allusions to a universal life energy are also occasionally referred to as the "occult forces of life." White and Krippner note that some individuals speculate that this life energy may be a "fifth force" or a "fifth type of energy" that interacts with the other four known forces of nature.[53] Woodhouse is one of those who propose that there may be a fifth force, which he refers to as "bioenergy."

Along these same lines is the commonly held belief of many alternative health practitioners that there is a kind of "subtle energy" which participates in human healing, especially nonlocal kinds of healing. Subtle energy also occasionally goes by the name of *ch'i* or *prana*. This fairly wide-held belief has generated a significant amount of controversy, especially over the last decade, because no empirical evidence has been found thus far to warrant the belief in this particular form of energy.

Distinctively relevant to the controversy surrounding the notion of subtle energy is physician Larry Dossey's discussion of energy in relation to consciousness and healing. In a number of articles he explains how the classical idea of energy does not fit with the new paradigm generated by new physics theories such as nonlocality.[54] His concern is chiefly focused on the nonlocal activities of consciousness that obviously participate in healing, but he also discusses para-phenomena which display "time-displacement," for example, precognition.[55]

Dossey notes that the idea of some kind of energy transmission is inappropriate for describing what is occurring during these events and therefore concludes that the whole idea of energy is incompatible with nonlocal phenomena. Rightfully so; there is no empirical evidence that there is any kind of energy transmission taking place.

Scientists are for the most part still thinking of energy as "force"— as a means of causing effects, rather than as an autonomous phenomen*on* that manifests itself into everything we see (and don't see). Contributing to the problem is that most of us are still searching for causal explanations. Most of us are still thinking and talking in classical, Newtonian terminology.

Dossey's reflections on energy, consciousness, and healing resonate with Jung's comments on the paradoxical nature of paraphenomenal events. Borrowing the phrasing of physicist Nick Herbert, Dossey notes that nonlocal events are considered to be "unmediated, unmitigated, and immediate."[56] For that reason, nonlocal physical phenomena, like the occurrence of split photons remaining somehow connected over long distances, seem to be acausal. Even when spatially separated, each photon seems to know what the other is doing and act accordingly; they seem to be parts of one whole rather than independent events.

Distant healing and other so-called paranormal events fit into the same category as the nonlocal phenomena known to occur in the microworld. There is no reason to believe that something *causes* them to occur. On this basis, Dossey proposes that we should discount the idea of energy as foundational to nonlocal healing:

> There appear...to be forms of distant healing in which energy appears to be not only undetectable but entirely absent. However, all forms of distant healing seem to have one thing in common: *they are related to consciousness.* We therefore would appear to be on safer ground to propose *consciousness instead of energy as the foundation for these events.*[57]

Dossey notes that local, or rather physiological, healing occurs in the classical cause-and-effect manner, where there is some kind of energy transmission or exchange, while nonlocal healing occurs somehow in relation to consciousness without energy transmissions, exchanges, or other physical causes. So it seems we are in a roundabout way back to the mind-body problem, albeit in a slightly more specific context related to ideas of energy.

Some individuals have criticized Dossey for proposing to discount "energy talk" when discussing nonlocal phenomena; however, perhaps it is best to discount energy talk, at least the kind of energy everyone is talking about. Although many of us claim to have departed from classical Newtonian thinking, the idea of energy as part of an enabling or disabling process — in essence, where causes end in effects and energy is used, transmitted, gained or lost — is still dominant. The idea that things are distinct, separate entities still dominates and pervades our efforts to understand everything. This idea is often still very much exemplified in the way practitioners speak of mind-over-matter healing, a topic we will turn to in the next chapter.

The Science of Unitary Human Beings

There are a number of nurse theorists who have developed conceptual models that provide a philosophical view on which nurses can base their practice. One of these models, developed by nurse educator Martha Rogers, is especially relevant here.[58] First published in 1970, Rogers' ideas are said to have initiated a shift in the paradigmatic view of nursing, and while many of her ideas reflect those of contemporary thinkers in other disciplines, she also holds notably unique positions that represent obvious departures.[59] Perhaps risking a bit of flak from a few of my colleagues, I will say that Rogers' model goes well beyond nursing.

Rogers affirmed that human beings are different from the sum of parts and different from matter, defining both unitary human beings and their environment as irreducible, indivisible, infinite, energy fields identified by pattern. Just as physicist Neils Bohr emphasized that an electron does not have a particle or wave-like nature, but rather that the electron is the wave-particle, Rogers emphasized that the energy field is the fundamental unit of all animate and inanimate things. Consequently, a person doesn't have an energy field; a person *is* an energy field. Our physical body is only one aspect of our individual human energy field.

The human and environmental fields are not believed to be subject to any causal laws. Instead, they are said to be in mutual process. Our changing nature is inherent and emerges from a mutual human and environmental field process. Therefore, there are no outside forces acting on us: all is integral.

According to Rogers the universe is entirely open and pandimensional—a nonlinear domain without any spatial or temporal attributes. In an open, pandimensional universe the concepts of time and space have little meaning. As Phillips notes: "Energy does not have to travel from here to there since both are together simultaneously anywhere in the universe; in a dimension transcending time and space—'there isn't any there.'"[60] And, because "there isn't any there," Rogers noted that her model offers a way of understanding various para-phenomena.

Rogers' idea of a pandimensional universe is in keeping with the mystic's perception of time. Psychologist Lawrence LeShan reflected:

> The separation in space between two events is an illusion and the separation in time between two events is an illusion. ...Sharp dividing lines are an illusion, and this includes the sharp dividing lines we customarily make between the past, the present and the future. Time, says the mystic, is "a seamless garment."[61]

In a pandimensional universe, past, present, and future merge into one, which Rogers referred to as the relative present.

Para-phenomena are considered quite normal in this view. And dying is a normal part of the human becoming process, in which the human energy field continues to exist after *physical* death. Indeed, Rogerians often say, "We don't die, we only change our cosmic address." Human beings are believed to be forever actualizing potentials in mutual process with the rest of the cosmos.

Rogers offered a radically different way of thinking and as such has inspired the development of many nursing theories for practice, one of which I shall introduce shortly. In Rogers' model energy cannot be transmitted or exchanged; everything is integral, and therefore causality is nonexistent. While her ideas come closest to conceptualizing a picture of the universe as a manifestation of the one phenome*non,* energy, there are some inconsistencies. Because everything is integral and change emerges from mutual process, outcomes are ultimately unpredictable, which is in keeping with quantum physics and with chaos and complexity theories. On the other hand, she so forcefully denied causality in any way, shape, or form that there is no room for any kind of teleology. Yet, some Rogerians have linked her ideas with those of Aristotle, Bohm, and Sheldrake, all of whom propose some kind of "that for the sake of which."[62]

A Different View

Somewhat in keeping with Rogers' view, but certainly not in keeping with many of the others, an explanation that is beautiful in its simplicity becomes apparent. Energy is everywhere. It can be intentionally changed or transformed. All things, both animate and inanimate, are manifestations of energy. So, too, are the more qualitative things, such as the fields within which there are said to be forces of gravity and electromagnetism, and the somewhat elusive neutrinos, along with all other particles that are sometimes not particles but waves. It seems that matter, in this instance the brain, is no different from the sentient abilities of living things; both the material of the brain and sentience are meaningful manifestations of energy, albeit different forms. All is essentially one.

The one is continually changing, transforming, and manifesting itself in different forms. Some forms are matter, and some are not, but as they all arise from one source, the multiplicity of energy forms are *connected* in such a way that there can be no mind-body dichotomy. Just as there is no void, no nothingness, there is no separation of mind and body since they are in essence forms of *one* thing. They are as one, different and yet complementary manifestations of energy. The form of energy that we call consciousness is distinct, although it may not be unique to human life.

Para-phenomena, in both the micro and macro worlds are simply the purposeful actualizing, changing, and transforming of energy. Individuals or things are not transferring energy anywhere. Rather, since energy is everywhere, it is being transformed or changed, intentionally and continuously. Nothing is standing still. In this view local and nonlocal events become two aspects of one thing, energy transformation. As such there is no difference between local and nonlocal occurrences other than that (we think) we understand and can account for the local events while we still cannot understand nor account for the nonlocal ones.

If we apply the quantum physics theories discussed earlier to human sentience, namely, uncertainty, nonlocality, and complementarity, which in fact many physicists have done, it becomes clear that all aspects of sentience must be integral with everything else. Consciousness complements but also goes beyond *physical* reality. Even the language we use, as Bohr, Heisenberg, and others have noted, is part of our sentient ability in the sense that the words and concepts we use help to develop ideas of *what is* and in so doing give us a greater awareness of reality. In the quantum microscopic world, everything is indivisible. How can we maintain that it is any different in the macrocosm?

Para-phenomena like psi, near-death experiences, and, indeed, synchronicities, aren't quite so surprising or difficult to accept when they are viewed within the framework of energy as a phenomenon. Viewed from this perspective they are indicative signs of a universal communal process. The phenomenon of energy is manifesting itself in human form, in a multiplicity of ways, and it is doing so purposively. So if abilities such as clairvoyance and telepathy are viewed as different but nevertheless meaningful transformations of the one phenome*non,* energy, it becomes easier to envision these kinds of abilities within the realm of every human being.

Studies of para-phenomena seem to suggest that everything is connected in an almost inexplicable way. Subjectivity and objectivity become merged within this framework; there is no dependency of human sentience on matter, just as matter is not dependent upon sentience. They are two aspects of *one* and the *same* thing; a manifestation of the *one* phenomenon, energy. And it seems that as we die — as we transform from what seems to be a tangible form to an intangible state of *still* being — we become acutely aware of the fact that *all is essentially one.*

Summary

Ever since human beings have developed the ability for thought there have been those who wonder "how" we do it. It has been a source of continuous motivation, for scientists and philosophers alike, to contemplate and investigate this phenomenon. There are a number of questions that arise whenever human sentience is examined: Is the mind and thus consciousness to be found in matter? Which came first, the matter or consciousness? Is there an everlasting soul? These questions, when looked at from the perspective of energy as a phenome*non,* take on different meaning.

The argument that the mind is in the matter is most profound in the neuroscience literature. However, there is also a large body of literature, particularly in philosophy, in which the converse argument is prevalent. That is to say, there are many who believe that the mind is not in the matter but rather is beyond the matter, or nonlocal. I propose an alternative view: all is essentially one, one that is in continual flux. However, the one is not physical or nonphysical; *it is energy.* All is essentially one thing, one phenome*non,* energy, which has manifested itself in a multiplicity of ways. Thus, it is not limited to *one form* although it is *one* thing, *one* phenome*non.*

I quote from scientist Erwin Schrodinger to stimulate further thought along these lines:

> It is the same elements that go to compose my mind and the world. This situation is the same for every mind and its world. ...The world is given to me only once, not one existing and one perceived. Subject and object are only one. The barrier between them cannot be said to have broken down as a result of recent experience in the physical sciences, for this barrier does not exist.[63]

Each of us is an alchemist, for we have the inherent ability to transform (or to change) different forms of energy as the need or intention arises. When I look outside my window, I see trees, among other things. But *do* I see trees — or do I see what I somehow choose to see? This is not a new idea. According to William James, "What we perceive comes as much from inside our heads as from outside."[64] My thoughts are actualities the existence of which only I can be entirely certain. And exist they do, for I can feel them. I hurt, I laugh, I cry, I am experiencing them.

When I hear music, I can feel it. When I read a book, I visualize the things I'm reading about. I don't respond to these things; I experience them. All of these to me are forms of action-at-a-distance. In this respect, the material aspects of the body are different from the sentient aspects of the human being. For neural action potentials are just that; they are synaptic energy processes which transform or change something within the physical body. They are limited to a certain place and time, whereas our mental abilities, although they appear to arise from within, lie not only within but also beyond the physicality of the body.

So it would seem that all aspects of our being — physical and otherwise — are still manifestations of one thing, energy. We have the innate ability not only to create reality in conjunction with the rest of the universe of energy, but by way of actualizing our inherent potentials, we can transcend the confines of *physicality.* There is purpose and directionality, although the direction certain aspects of our being takes (here I am referring to sentience and psi), must be nonlinear, as it is nonspatial and nontemporal. While we ultimately cannot choose not to age or not to have the physical body deteriorate to the point of extinction, our sentient abilities can move forward, or backward, as we choose.

We might ask ourselves why it is that people who have had a NDE often go on to lead more spiritual lives. Why is it that they turn outward, rather than inward, and focus on helping others? It does seem that there is a reason for it all, some kind of universal purpose. Then there are our subliminal abilities. Dreams, intuition, and synchronicities. Being in the right place at the right time. How is it that a dream or a meaningful coincidence can change the direction of our lives? Why do these things occur? Unknowingly, it seems, we are participating in some universal purposive communal process. It is a communal process rather than a causal one in the sense that it, the *one,* is changing, purposively actualizing inherent potentials. So, not surprisingly, we sometimes make choices that unknowingly, but still purposefully, contribute to the larger whole.

Recently my husband and I had a brief sojourn at Lake Tahoe. The day we were leaving and had in fact already driven about ten miles, I had, for some reason, the desire to turn back to the lake to take more pictures of the snowcapped mountains. I didn't know why; we had already taken plenty of pictures. But we drove back alongside the lake and found a park-like setting to pull into. My husband parked the car, and we walked along the lake.

A short distance from where we parked was a wooden staircase leading up to a deck overlooking a very rocky beach below. We had barely made it to the top, when my husband cried out and jumped across the deck to get hold of a little boy, perhaps four years old, who was just about to go crashing down on the rocks below. He had been standing on a bench and was leaning precariously over the rail. His guardian was on the beach and seemed unaware of the danger. Was it chance that brought us there to that spot on the lake, at that point in time? I don't think so. I think about that little boy all the time and wonder what he will contribute to the world when he grows up. For surely that day we had, on some subliminal level, engaged in a purposeful universal communal process.

Most of the time we go through life without this kind of awareness. We don't see the connections. Synchronicities and psi are apparently not at all uncommon; yet many people don't attach any meaning to them. Accordingly, many of us don't take notice when perhaps we should. But I for one am convinced that the boundaries we think exist do not. All is essentially one, although nothing stands still. Nothing is ever lost, just changed. We are truly engaged in an eternally changing purposive dance with the rest of the cosmos.

Chapter 5

The Nature of Health and Healing

> Shall we begin by taking it as a general principle — that *all disease,* at some period or other of its course, is more or less a reparative process, not necessarily accompanied with suffering.
>
> — Florence Nightingale[1]

Why is it that living things are so similar and yet so dissimilar? We all have unique but nevertheless comparable hearts, lungs, and fingerprints. Why is it that without any seeming intent or effort on our part, most of us manage to be healthy? In keeping with the becoming of it all, underlying all aspects of health and healing are signs of *unity* and *purpose.* Just as we are connected to the stars above and the earth beneath our feet, just as every snowflake, although unique, has a discernible pattern, so too does every human being.

Health is an enigma, as philosopher Hans-Georg Gadamer notes: "The real mystery lies in the hidden character of health. Health does not actually present itself to us.... It lies in the nature of health that it sustains its own proper balance and proportion."[2] I will say that the nature of health and healing *is* a mystery because it is irrevocably linked to the enigma of energy. Health and healing are yet other examples of the purposive nature of the *one.*

Conventional Western medicine, however, has developed in accordance with Newtonian mechanics and the Cartesian belief that the mind and body are separate entities. Traditionally we have been regarded as equivalent to machines; our broken parts can be fixed or replaced if only we have the knowledge to do so. Moreover, a person *could* be reduced to parts because each part has been perceived as a separate and distinct entity.

Certainly great strides in the cure and prevention of disease have been made; the increasing longevity of human beings is a strong indication that medical science has been very successful. However, mechanism (reductionism) does not fit into the realm of modern health care. Perhaps it belongs in the laboratories of physiologists who must necessarily look at parts in order to comprehend their functions. But even there it may be of limited value.

In professional practice we now know that there are limits to what can be accomplished by focusing on the parts rather than the whole. I say this in a very practical sense, for I've seen many patients who were "cured" of the ailment that led to their hospitalization, only to develop other illnesses or problems. In health care they are often called complications, but on a closer look, such complications are often a result of ignoring the whole person and focusing on just one body part.

Some years ago I was working as a clinical nurse specialist in a teaching hospital. While I was on a unit I happened to notice a call bell light on for one of the patient rooms. I walked in and found an elderly lady sitting in a chair by the window. When I asked if there was anything she needed, she replied, "Yes, I'd like to know when my treatment is going to start. I don't have too many more sunny days left, you know." She looked frail, quite ill really, and I suspected right away that she was talking about chemotherapy or radiation treatments. When I questioned her further, I found that she was indeed asking why her chemotherapy hadn't been started since she had already been in the hospital for three days.

Truly there was no acceptable reason. She had breast cancer which had spread to her bones and was to receive chemotherapy as a palliative means of making her more comfortable. The patient was dying, and she didn't want to waste the time she had left here on earth in the hospital with strangers who didn't seem to care one wit about her. I have frequently seen this kind of thing happen, and it is because no one is thinking about the person, the whole.

I have always thought that the key to being a good health care professional is to think of each patient as one of your own. At least that way you can't forget that you're caring for a human being. And I often find myself advocating for very basic things, things for which no one *should* have to advocate. For instance, a number of years ago I had a patient who was being worked up for cardiac disease. The physician requested numerous tests but never noticed that the patient's blood sugar was elevated. So busy was she, focusing on the patient's heart, that she never noticed he was diabetic. It turned out that the patient's elevated blood sugar was the real problem. His heart was fine.

If everyone involved in the care of patients thought of them as family members, issues like the timely initiation of chemotherapy and the proper evaluation of all pertinent lab results wouldn't be problems. The providers of care would be more sensitive to the fact that they are caring for human beings, *real* persons who think, feel, and dream, persons who cannot be broken down into parts in order to understand the whole.

Recently, while I was working with a local bereavement group, one of the members asked me why it is that after someone dies in a hospital, all the family members ever receive from the hospital are bills. She asked why the hospital doesn't send a sympathy card at least, before they start bombarding the family with bills, which often bounce around for at least a year after someone has died. It usually takes at least that long to get everything settled. I couldn't answer her, of course, but I wholeheartedly agreed with her; the families of our patients *are* our patients too. Nothing is separate.

Although it seems that we compartmentalize and try to reduce everything to separate parts in today's health care world, it hasn't always been that way. While there is real disparity between the eras in which they lived, Hippocrates (approximately 460–370 B.C.) and Nightingale (1820–1910), founders of medicine and nursing respectively, shared similar views concerning the nature of health and healing. Both had an intuitive comprehension of the natural purposive manner in which human beings strive toward the attainment of health. And they looked at wholes. Thankfully, though, we are moving toward a unitary perspective in health care, which is especially evident in Rogers' science of unitary human beings. In a strange way this movement seems similar to the wholeness approach apparent in Hippocrates' and Nightingale's work.

On the next pages we will examine various ideas related to the nature of health and healing. Often called models or theories, just as in other scientific disciplines, the different health care models lead to ambiguous and sometimes conflicting ideas. But the disparities arising from these different models are not the focus here, and I refer the interested reader to other professional writings.

What I hope to illuminate is that health and healing are manifestations of purposive energy transformation. Wholeness in this respect amounts to *one* thing, which has manifested itself as many distinctly unique things — us — and is purposefully changing to actualize the human potentials of health and healing.

A Healing Force of Nature

Medicine existed in a primitive manner, largely directed by spiritual belief and superstition, long before the Grecian era. Very little is known about primitive medical folklore and the practices in early cultures; however, it is interesting to note that there was a common belief in supernatural *forces*. The ancient Greeks, for the most part, were convinced that there were elementary forces or qualities in the macrocosm (universe) that went hand in hand with the basic elements found in nature.[3]

Hippocrates noted a relation between the human being and the universe: the human is a smaller but still distinct form of all else in nature. He believed that the body consisted of four basic humors (substances) — blood, phlegm, yellow bile, and black bile — and further identified that these humors had qualities such as heat, cold, dryness, and moistness. The four humors were said to reflect the elements earth, air, fire, and water. One was considered healthy if there was a proper balancing of the four humors and their qualities, while conversely any imbalance or excess resulted in illness.

Recognizing the value of proper nutrition and the uniqueness of individual needs during times of illness, Hippocrates emphasized that people not only differ in their normal nutritional requirements, but also when they are sick, their needs change. He stressed that environmental conditions play a part in both health and illness and urged practicing physicians to familiarize themselves with the geographic conditions where people lived.

One cannot help but feel a sense of awe when reading the Hippocratic works. Much of his work shows exceptionally keen insight concerning health and illness. One particularly striking example is to be found in the Hippocratic writings on dreams, where it is suggested that prayer be used as a means to promote healing. Hippocrates didn't advocate using prayer as the sole resource for healing; however, he did include prayers in the treatment of various ills. This has particular relevance since prayer has resurfaced as a potential healing modality over the last few decades.[4]

The most important observation Hippocrates made was that the human being has an innate capacity to heal:

> The body's nature is the physician in disease. Nature finds the way for herself, not from thought. For example, blinking, and the tongue offers its assistance, and all similar things. Well trained, readily and without instruction, nature does what is needed. Tears, moisture of the nostrils, sneezing, ear wax, production of saliva in the mouth, the intake of breath, exhalation, yawning, coughing.[5]

This particular statement formed the basis for the Hippocratic *vis medicatrix naturae*—a force of nature that cures from within.

Underlying this healing force is the notion of equilibrium, a balancing of energy. Historian of medicine B. Inglis likened the *vis medicatrix naturae* to a life force that encompasses all living things, noting that its basis came from evolutionary processes:

> To preserve nature's balance, evolution had also to provide some counter to disease; and the main instrument which it perfected was the built-in recuperative power which the Hippocratic writings called *vis medicatrix naturae;* the life force. "Life force" is not an ideal translation, as colloquially it can embrace anything from the act of breathing to an abstraction like the "will to live"; but in a medical context it is usually applied to those measures which are taken automatically in life, even down to the earthworm level, to repair damage, or to throw off the effects of poisons. A cut will heal, if left alone; a broken bone will knit together again.[6]

The founder of modern nursing, Florence Nightingale, also noted an innate healing ability and made the astute observation that "the reparative process which Nature has instituted and which we call disease has been hindered by some want of knowledge or attention."[7] And so, like Hippocrates, Nightingale emphasized the importance of environmental factors in relation to human well-being, advocating clean rooms, adequate ventilation, nutrition, light, and warmth, among other things.

It is clear when one scrutinizes her work that an idea of energy is ubiquitous. For instance, she emphasized the importance of sunlight, having gleaned this from her observations:

> It is a curious thing to observe how almost all patients lie with their faces turned to the light, exactly as plants always make their way towards the light; a patient will even complain that it gives him pain "lying on that side." "Then why do you lie on that side?" He does not know, — but we do. It is because it is the side towards the window. A fashionable physician has recently published in a government report that he always turns his patient's faces from the light. Yes, but nature is stronger than fashionable physicians, and depend upon it she turns the faces back and towards such light.[8]

Sunlight is considered to be a direct source of energy for all living things. Plants involuntarily, or so it seems, lean toward sunlight, as if they want to capture every bit of energy possible. So, too, do ill persons who seem to unconsciously reach out for the energy source that only sunlight can give. But, I wonder, are they reaching for an energy source? Perhaps they are unconsciously engaged in transforming the universal phenomenon of energy and hence are actualizing their unique healing potential.

Nightingale believed that providing variety in a patient's surroundings, most especially things of beauty and of color, were conducive to well-being, and that these amenities had historically been neglected by health care providers. What's more, she noted that patient's cravings and "fancies" were usually exactly what the patient needed to recover; the patient often knew better than health care providers what was needed to regain health. Like Hippocrates she found that certain foods which would not be beneficial to a healthy individual were beneficial to the ill and that the sick frequently were the ones who intuited this without actually knowing it.

Even spirituality, as portrayed by Nightingale, seems to be an energy form, a potent form capable of assisting the healing process.[9] According to her, God made all the physical laws in our universe, and she firmly contended that healing is a natural process that is also governed by laws. Hence, her frequently quoted words:

> It is often thought that medicine is the curative process. It is no such thing; medicine is the surgery of functions, as surgery proper is that of limbs and organs. Neither can do anything but remove obstructions; neither can cure; nature alone cures. Surgery removes the bullet out of the limb, which is an obstruction to cure, but nature heals the wound. So it is with medicine; the function of an organ becomes obstructed; medicine so far as we know, assists nature to remove the obstruction, but does nothing more. And what nursing has to do in either case, is to put the patient in the best condition for nature to act upon him.[10]

On a similar note traditional modern-day physicians and nurses allude to this innate healing force using terms such as "homeostasis," meaning, "the staying power of the body," and "energy conservation."[11] Alternative health care practitioners also recognize this purposive healing nature. Andrew Weil, a physician and author of several books about health and healing, forcefully puts forward the idea that the body can heal itself. He notes a number of principles that are held to be true by many alternative health practitioners:

> The body wants to be healthy. . . . Healing is a natural power. . . . The body is a whole, and all of its parts are connected. . . . There is no separation of mind and body. . . . The beliefs of practitioners strongly influence the healing powers of patients.[12]

Although traditional Western medicine continues to dominate health care in America, alternative (also known as integrative, complementary) healing modalities have become better known and are sought by and helpful to many people. But, in fact, the basic ideas underlying alternative health care have existed in almost every culture since time immemorial, and many of the alternative modalities practiced today were derived from ancient Eastern practices.

In the West, for the most part, alternative health care has been considered nonscientific and thus has not been given the attention its success surely deserves. Caplan notes, "At one time or another, the AMA has denounced homeopathy, osteopathy, optometry, acupuncture, self-care, chiropractic, midwifery, and lay analysis as being either dangerous, or fraudulent, or both."[13] It would appear that, as Inglis puts it: "The most serious error . . . into which orthodox medicine has fallen is too narrowly to restrict the confines of science. The tendency has been to insist that whatever is not seen or not understood does not exist."[14] But times are changing. Many physicians, nurses, and psychologists are learning, teaching, and practicing alternative healing modalities.

The problem with alternative modalities is that many of them cannot be explained by the traditional cause-and-effect thesis. So the new trend has been to link these alternative therapies with physical theories such as electromagnetic field theory, nonlocality, complementarity, and others. These theories are believed to offer a framework from which the alternative health care practitioner can try to account for the evident success of healing modalities such as imagery, prayer, and therapeutic touch.

The Human Energy Field and Energy Therapies

Alternative healing modalities are often called energy therapies. Nurse Victoria Slater writes about energetic healing (EH), which she says is a metaphoric term referring to "healing occurring at the quantum and electromagnetic levels of a person, plant or animal." She discusses three quantum physics theories in conjunction with energy therapies: potentiality/actuality, observer-created reality, and consciousness-created reality.

The potentiality-actuality theory that Slater discusses refers to the wave function (potentiality) which collapses into a particle (actuality):

> Practitioners of virtually any discipline will agree that all actualities exist, first, as potentialities. Until a wave "collapses" into a particle, only potentiality exists; once the wave collapses, a particle's actual physical characteristics, such as location and velocity, become manifested.[15]

This account of the quantum wave-particle theory is obviously not consistent with Aristotle's concept of *energeia* because, according to him, actualities must necessarily precede all potentialities. In other words, nothing can come from nothing.[16]

Once again, we get into the subject-object dichotomy, which, after all, doesn't really exist. This is where the theories of observer-created reality and consciousness-created reality are relevant. Here we re-encounter Heisenberg's uncertainty principle and Bohr's complementarity theory. It is well known that in the quantum world we cannot accurately measure various things simultaneously without concurrently changing them and, hence, changing our reality. Energy therapies seem to share the same characteristics; thus, traditional cause-and-effect relations simply cannot explain them.

Many alternative health practitioners believe that they assist the patient by eliciting the support of some kind of vital energy or force that extends beyond the realm of the physical body. As noted earlier, the idea of a vital energy is implicit in many Asian cultures, particularly in the form of *ch'i*, *qi*, and *prana*. These terms are becoming increasingly popular in American culture. Under the auspices of alternative modalities there have been numerous terms used to describe this vital energy that is said to be a part of the healing process, words such as "subtle energy" and "bioenergy." Examples are numerous; however, a few should suffice to bring home a very important point. The practice of homeopathy is said to "connect with the spiritual vital force," while acupuncture is said to "utilize *qi*," and "Ayurvedic medicine and yoga teachers are in touch with *prana*."[17]

It is obvious that this energy, no matter by what name it is called, is not just within the body; it exists within and without. It is a cosmic, all-pervading energy. In addition, many alternative health care professionals work with the human energy field believing it to be the most fundamental aspect of the human being.[18] Some practitioners note that there are several fields rather than one. Whether it is posited that there is one or many, the general contention is that these energy fields are integral and infinite with the universe.

Perhaps partly because of these beliefs, many alternative health care practitioners consider healing to be of a spiritual nature. Health care practitioner Brian Seaward notes: "What the mechanistic medical paradigm failed to incorporate was the dimension of the human spirit, an unmeasurable source of energy with a potential healing power all its own."[19] In a similar fashion, Slater notes that "all healing . . . is ultimately spiritual healing."[20] Spirituality may or may not be identified with a specific religion. It is primarily a way of viewing the world which emphasizes the immaterial or incorporeal aspects of nature. In alternative health care, spirituality is markedly identifiable in the way the mind is believed to play the most vital role in the actualizing of the human potential for healing.

Approaches to healing that fall under energy therapies include meditation, homeopathy, imagery, biofeedback, and hand-mediated modalities, among others. Hand-mediated energy healing modalities include healing touch, therapeutic touch, polarity, reflexology, and acupressure. Several of these hand-mediated modalities cannot be explained within the traditional medical framework because in many instances actual touch is not a necessary correlate for healing. There is no observable means of quantifying the mode in which these therapies work; there is no empirical evidence that energy is being exchanged or transferred. The only way we can measure their "effects" is by observing the outcomes of the therapy, that is, by seeing if healing has occurred. Yet, there is a growing body of research that supports the effectiveness of some of these alternative healing modalities.

Therapeutic touch (TT) is one alternative modality that is often said to work by means of some kind of energy transfer or transmission, although actual touch isn't involved.[21] Of course the skeptics, and there are plenty, think it's quackery because there is no empirical evidence that energy is being transmitted. Yet, there is a growing body of research that supports the *effectiveness* (again a causal term) of TT.[22] What I question is the "energy talk" surrounding TT, which implies that there is some kind of energy exchange or transmission going on between the healer and healee. Shortly I will offer my own view.

Many alternative care practitioners discuss theories that might account for the apparent orderliness of health and healing, such as the implicate order postulated by Bohm and the morphogenic field theory proposed by Sheldrake. Talbot suggests that the human energy field may "function as a kind of blueprint from which the body gets its structural cues."[23] Some, such as Slater, discuss possible reasons for the underlying order of human energy by linking the human energy field not only with Sheldrake's theory of morphogenic fields but also with the Eastern views of the aura and chakras.[24] Again, we have indirect references to some kind of formal and final causation.

Spontaneous Healing

The phenomena of spontaneous healing and remission seem to exemplify a kind of universal purpose. Unfortunately, as with alternative healing modalities, a lot of skepticism still surrounds these occurrences. Brendan O'Regan, vice president for research at the Institute of Noetic Sciences, notes that spontaneous healing is frequently believed to be a case of misdiagnosis, even after numerous studies have confirmed that it wasn't. I found particularly interesting the definition O'Regan provided for the term "spontaneous":

> "Spontaneous" means "acting in accordance with, or resulting from feeling, temperament or disposition, or from a native internal proneness, readiness or tendency without compulsion, constraint or premeditation." Another meaning is "acting by its own impulse, energy, or natural law without external cause or influence."[25]

While "spontaneous" healing implies a total cure, "remission" means that the disease is lessening. There are no physical scientific explanations for cases of spontaneous healing or remission, and while it has yet to be verified by scientific investigation (in the strict empirical sense), it is becoming increasingly more evident that this kind of healing does occur.

Certain patterns seem to repeat themselves in the reported cases of spontaneous healing and remission. One of the most interesting and relevant for discussion here is that many cases of remission are preceded by an infection of some sort. O'Regan reports that there are documented cases of cancer remission after bacterial infection in patients with cancer as far back as the 1890s. Dr. William Coley of New York

> noticed . . . that his patients who became infected with a bacterial skin infection . . . would react with a fever and spend a few days fighting it off — as though the immune system were being activated to fight off the infection. Then in about 40 percent of the patients, the cancer would disappear.[26]

Since then, there have been other findings confirming that a pattern exists between the onset of infection and remissions of various diseases.

That infections often precede remissions of terminal illness illustrates the fact that human energy changes in a purposive manner; it actualizes its inherent potential to heal. And the knowledge that infections play a part in cases of remission suggests the possibility that in health care we are going about healing in the wrong way. All diseases may be another means by which we actualize human potential. Although it is surely hard to understand why people suffer from diseases like cancer and AIDS, having myself lost many family members, friends, and patients to these and other horrible diseases, it may well be that disease is another means by which we actualize our innate potentials, just as dying is a means of transcending the physicality of the body.

On the pediatric unit of a hospital, the most vibrant, vivacious children are usually the ones who have chronic, and sometimes terminal, illnesses. You can always pick them out. They may or may not look ill, but there are other characteristics they always seem to display, like compassion for the nursing and medical staff, as well as the housekeepers and the other children on the unit. They spend their time in the hospital looking outward toward others rather than inward. They are actualizing their unique potentials.

Magic Johnson, the renowned N.B.A. basketball player, is an example of someone who, in the face of potentially life-threatening illness, is actualizing potential. He has forever changed the course of basketball history with his agile and strategic moves. In 1991 he tested positive for HIV, the precursor for AIDS. Johnson has been involved not only in basketball, but also in an impressive array of causes for the betterment of society, including the US National AIDS Committee, the United Negro College Fund, as well as the American Heart and Muscular Dystrophy Associations.

Thankfully, Johnson continues to be healthy and to work his magic in many ways on society today. Yet it seems that he has actualized innate potentials, which may have stayed forever dormant, had he not tested positive for HIV. In this regard I recall the traditional saying "that God doesn't give us more than we can handle." And I also recall that individuals who experience NDEs go on to lead more spiritual lives, often turning their focus outward toward others. Surely, Magic Johnson and others like him portray this kind of spirituality. Without doubt, their personal misfortunes *have* made the world a better place. And I can't help wondering how many youngsters today practice safe sex, how many people do *not* have AIDS, because Magic Johnson reached out to us, sharing and educating the public.

Miracle Healing

Miracle healing, sometimes called spiritual healing, is a form of spontaneous remission and healing that is thought to be divinely or spiritually related. What sets miracle healing apart from spontaneous healing in general is that it frequently seems to occur without any effort on the part of the person who experiences the healing. Dossey notes:

> One of their chief characteristics is that they do not determinedly want healing; they are not desperate for a miracle to occur; they are not trying at any cost to extract a radical healing from the universe. They have a quality of acceptance and gratitude, as if things are quite all right in spite of the presence of the disease.[27]

It seems that there is some reason underlying the miracle healing or at the very least some organizing principle at work. Perhaps the individuals who experience miracle healing do so in order to complete some universal purpose.

Obviously, miracle healing falls within the discipline of theology and will not be discussed in great detail here; however, it is interesting to note that this type of healing is reported much more frequently than one would think. For instance, spontaneous remissions or healings have been experienced at Medjugorje in the former Yugoslavia, now Croatia, where apparitions of the Virgin Mary are reported to have occurred. O'Regan notes that reports of these cases are similar to the reports of healing that continue to occur in Lourdes, France, ever since 1858 when an apparition of the Virgin Mary occurred.[28]

Participating Knowingly in Change

Rogers' model offers a different perspective on health and healing, yet her view also reflects the ideas of unity and purpose.[29] In a pandimensional universe, past, present, and future merge into one — the relative present — so that when an elderly person mixes past with present, it is not necessarily pathological; it may well be that the person *is* experiencing the past in the present. Spontaneous healing and remission would surely be characterized by Rogers as a mutual process of innovative becoming, where, since the universe is pandimensional, human beings are able to participate knowingly in change in ways that most of us normally, in our three dimensional view, cannot accept nor comprehend.

Within a Rogerian framework all experiences, including those during times of illness or crisis, are viewed as the actualizing of human potential rather than abnormal or deleterious. When a dying person says that he has seen or talked with a dead loved one, nurses and physicians who hold a Rogerian unitary view don't jump to the conclusion that the person is hallucinating, nor do they attribute these para-phenomenal experiences to the administration of pain medications.

I have been with many patients and their families during the process of dying. Many patients who are dying begin to see and say things that most of us would think are abnormal, and frequently they claim to communicate with dead loved ones. In a pandimensional universe, where the human environmental fields are infinitely becoming, these kinds of experiences not only make sense, they are quite normal.[30]

To give a practical everyday example of Rogers' ideas on health and healing, we can look at organ transplantation. Commonly accepted generalities based on causal relationships cannot justify why certain patients reject the donor organ and others do not, even though the immunosuppressive therapy they all receive may be the same.[31] In fact, no one can account for the obvious success of many transplants when it seems that the tissue match was poor. But statistical data point to a correlation we cannot deny. Some patients, despite the fact that the odds are against them, do well, while others who seem to have everything in their favor, do not.

Joe was forty-three years old, with a wife and two young children. He had developed cardiomyopathy, a disease of the heart muscle that eventually leads to death. He was critically ill on his admission into the hospital and had been moved up to the number one slot on the waiting list for a heart transplant. A call came through early into the evening shift; a donor had been located. A young woman, Susan, had been in a terrible car accident and was declared brain dead. She had chosen to donate her vital organs as well as her eyes. The tissue match was poor, but it was the only chance Joe had since the doctors thought he wouldn't live much longer.

In Rogers' view the transplant recipient, Joe, and donor, Susan, are knowing participants in a mutual process of becoming; both individuals are becoming increasingly diverse and innovative. In cases such as this, where there is vital organ donation, for instance, a heart or liver, the person, despite imminent physical death, is still becoming increasingly diverse and moving toward a state that Rogers would call beyond waking. In this case, Susan may change her cosmic address, but she is still in the process of becoming. This mutual process will be unique for every individual. Because of this, outcomes are ultimately unpredictable; we don't know which transplants will be successful and which will not. However, health care professionals knowingly participate in the healing process by assisting the patients to actualize their unique potentials — whatever those potentials may be.

It has been noted that there is an increasing frequency and vicissitude in the incidence of diseases that result in organ failure. This is in keeping with Rogers' belief that we are moving toward higher and higher frequencies. A good example is the augmenting frequency of cardiac disorders, particularly cardiomyopathies of unknown origin such as Joe's, a disease where the heart pump becomes incapable of maintaining adequate circulation. Although there appears to be a relation between lifestyle and the incidence of various cardiac diseases, researchers are constantly struggling with contradictory findings, so much so that they have eliminated the term "proven" from their repertoire and now admit that nothing is entirely predictable. What medical science once accepted as causal in nature no longer rings true. This is consistent with Rogers' view of an acausal universe.

Incorporating Rogers' fundamental ideas into a comprehensive framework, nurse, psychotherapist, and educator Elizabeth Barrett developed a theory of power, proposing that "power is the capacity to participate knowingly in the nature of change . . . as manifest by awareness, choices, freedom to act intentionally, and involvement in creating changes."[32] In a process called health patterning, many nurses use the theory of power to assist patients in the actualizing of their unique potentials.

Health patterning entails the mutual participation of health care practitioner and patient in dialogue as well as other healing modalities such as imagery and therapeutic touch. Patients come to know their choices and make informed decisions to facilitate well-being.

Because this process is a mutual, rather than causal one between the human being and his or her environment, outcomes are said to be unpredictable.

You might ask, why bother with it if you can't predict the outcome? Why bother with something that can't be empirically proven to work? Well, both Joe and Susan had been aware of their choices and had intentionally chosen to be recipient and donor. Neither knew whether the transplant would be successful or not, but they both chose to be involved in creating change. And not surprisingly, at least from a Rogerian view, Joe is doing well despite the fact that the tissue match was poor.

Many patients who participate in health patterning return to their health care practitioners on a regular basis. Patients experience a powerful sense of freedom when they let go of outcomes. They will tell you that knowing participation in change — being aware of the choices one can make and being actively involved in their health care activities — makes a profound difference in their feelings of health and well-being even though outcomes are ultimately unpredictable. And when we knowingly participate in change, we are intentionally participating in the universal, communal process of energy transformation.

Although the revolutionary ideas of Rogers have been, to date, largely contained within the discipline of nursing, her synthesis of ideas has created a new way of viewing the universe, human health, and healing. And there are practitioners in other health care disciplines that are becoming familiar with her views. Physician Steven Field notes that Rogers' unitary view offers a way of conceptualizing what it is physicians do, stating that "the art of medical practice not only can be done, but in fact exists within a Rogerian framework."[33]

While Rogers' view comes closest to the idea that all is essentially one, I can't seem to let go of the idea that there is a "that for the sake of which" in everything, and, unfortunately, Rogers strictly ruled out final causes. But purpose does not imply cause and effect; it implies meaning and intention. Of course, it may be that I'm viewing things in that three-dimensional manner Margenau referred to, but nevertheless the teleological nature of health and healing seems so obvious to me that I simply can't attribute it to the idea that we *always participate knowingly* in change.

Another View: Transforming the One

Since the human mind can display an inordinate degree of power, many alternative health practitioners believe the mind can heal the body. Often termed "mind-over-matter healing" or "mind-body healing," this kind of healing is also discussed in terms of cause-and-effect relations. Terms like "affects" and "mind/body interactions" are frequently used so that we are given the impression that there are two things interacting in a cause-and-effect manner. One example of so-called mind-over-matter or mind-body healing that is currently gaining widespread publicity is prayer. Within the last twenty or thirty years, an increasing body of documentation regarding prayer healing has been published.[34] Dossey notes that it cannot be scientifically explained, since prayer healing occurs nonlocally, and, in fact, sometimes the person praying is a great distance from the individual he or she is praying for, and yet the person heals.

We touched upon this earlier, and it is obvious that the idea of energy as part of a causal process is not applicable for alternative modalities such as prayer healing and therapeutic touch. So, appealing to Ockham's razor once again and working on the premises of the previous chapters, I suggest again that mind and body are not *two* things but rather different aspects or forms of one thing — energy. There is no separation, so instead of the mind acting on the body, the mind and body are simultaneously actualizing potentials. From this perspective we can begin to view health and healing in a different light.

To put this in a proper perspective I can give a personal example. Often I experience the ergonomic aches and pains that have been attributed to sitting at a computer for a long time. Sometimes these pains get so distracting I am forced to stop my work. I then use a technique I had found in an article online called "Actualism" written by spiritual teacher Bruce Jaffe.[35] It takes about ten minutes, and then the aches and pains are gone. What I am actually doing, I think, is transforming myself (an energy form) along with the rest of the universal energy into a means of actualizing the potential for "no pain." I am transforming or changing myself in a manner conducive to comfort rather than pain. Exercises like this one are becoming increasingly common in the United States, although in the Eastern countries they have been practiced for centuries.

I do not say I am transmitting, transferring, or channeling energy. I don't say that because if it is all one, where exactly could I be transmitting the one to? It's everywhere, so that I can only be changing the one. There are no separate parts to exchange energy with because everything is a manifestation of the *one.* Viewing energy as a phenomenon means a radical change in the way we think and speak. Since I am energy, it makes no sense to say, or even think, that I'm losing or gaining energy, because if I am, I am losing or gaining parts of me.

There are obviously certain energy transformations that are necessary. For example, I couldn't go without eating or sleeping or drinking for any length of time without my physical body transforming into a totally different body and, eventually, a very different form of energy (I'd have no *body*). A rosebush transforms sunlight and rainwater into budding, poignant roses; without these transformations of energy a rose would never bloom and we would never be able to see or smell it.

Thinking of energy as a phenomenon offers a possible explanation for why, when there is no discernible reason, we seem to have increased vitality. A man who stays up all night awaiting the birth of his child and then goes straight to work or runs around all day handing out cigars doesn't seem to suffer any negative consequences from his lack of sleep. He doesn't walk around like a zombie; he seems to have what we think of as increased energy. The woman who is laboring with the birth of their child keeps going, often despite lack of food or rest. But how is it possible? Where did the increased energy *come* from? In fact, it seems they have transformed the universal phenomenon of energy, of which they are both an inseparable part, in a positive way to actualize their potentials for increased vitality.

In the writings of philosopher William James there is evidence of this idea. He referred to the occurrence of increased vitality when there seems to be no explanation for it as "second wind," noting that we sometimes go beyond what we consider the *normal* range of human ability. It seems that we can push ourselves to transcend feelings of fatigue without suffering any permanent negative changes. James contended that we have energies over and above what is needed in order simply to live: "But the plain fact remains that men the world over possess amounts of resources, which only very exceptional individuals push to their extremes of use."[36]

Yet it seems clear that we do not have excess stores of energy; along with a multiplicity of other energy forms, we *are* explicitly a manifestation of energy. None of it is separate. It is all so intricately connected that in essence there is only one whole. We could say that it's the way we actualize the potentials of energy, the way we transform, that sometimes results in the *illusion* that there is an excess store of energy.

Despite the fact that health and healing are in so many ways enigmas, we have continued to try to quantify them according to Newtonian mechanics. We continue to posit energy exchange and cause and effect as the means by which healing occurs. We posit that a different kind of subtle energy exists, in order to retain our mechanistic view. Many practitioners reflect the idea that energy is somehow being balanced, or exchanged, as nurse Janet Macrae's comments illustrate:

> A human being is . . . a localized interaction pattern within various universal fields of energy. . . . Body, mind, and spirit, therefore, are not separate substances or categories, but rather different energy frequencies that are continually interacting. A human being is a complex, multidimensional energy system. From this point of view, health or wholeness implies an inner balance among these different levels or dimensions of energy as well as an open, harmonious interchange between the individual and the environment.[37]

Although her words impart the idea that energy is somehow being transmitted, interchanged, or exchanged, she also gives the impression that all things are manifestations of energy. The problem is that if everything is essentially a manifestation of one thing — energy — how is it being transmitted or interchanged *from* the one *to* the one? How is the one *interacting* with itself?

I will suggest that it is easier to understand how alternative modalities are successful if we think of energy as a phenomenon that is always purposively changing. We participate in the healing process. We actualize our healing potentials, and with conscious intention (another form of energy) we can often direct energy transformation. It's likely that the reason why we can't quantify many of these alternative healing modalities and explain them in classical Newtonian terminology — cause and effect, energy exchange — is because there is no cause and effect or energy exchange. Rather, the pheno*menon* of energy — the one — is transforming; however, not only is it transforming, it is doing so *purposively.*

The *one* is purposively becoming whatever it's meant to become. It's not always so obvious, nor does it always seem fair. Take, for example, Stephen Hawking, the most brilliant theoretical physicist of this century. Hawking was diagnosed many years ago with Lou Gehrig's disease, a rare and crippling neurological disorder. At the time he was told he had only a few years left to live, but today, decades later, he is alive and has done miraculously well.

Hawking has forever changed the course of theoretical physics, accomplishing more while severely physically debilitated than most other people accomplish in a lifetime. Many physicians claim it is a miracle that Hawking is still alive. But many of his colleagues profess that Hawking's intellect and memory are the real miracles. John Boslough, author of *Stephen Hawking's Universe,* discusses Hawking's work and his physical disability, which he claims, put Hawking in an advantageous position to develop his theories. The striking idea behind all this seems to be that Hawking's mind, his supreme intellect, has developed into such proportions somehow because of his debilitating physical condition.

Instead of positing a cause-and-effect explanation, we could say that Hawking developed one form of energy, or rather transformed one form of energy, namely, his intellect, in lieu of, or at least over and above, another form, namely, his physical body. This wasn't intentional on his part; I certainly don't think he would have chosen to be wheelchair-bound as he has been these many years. But I do think he exemplifies some universal purpose, some kind of perhaps *unknowing participation in change.*

It seems to me that individuals like Stephen Hawking, Magic Johnson, Helen Keller, and many noteworthy others exemplify a universal purpose. Without them, this world would be a very different place. And, were they different, their unique and truly significant contributions to humanity would surely be different as well. What we perceive as human illness or handicap may very well be another means by which energy transforms purposively, actualizing its potentials for the well-being of the *one.*

Summary

While in many respects the traditional, alternative, and Rogerian models are markedly different from one another, there is a common thread, a basic philosophical underpinning that these ideas of health and healing share. Everything is intricately connected, and whether we posit that there is an energy exchange or an energy transformation, the processes involved in health and healing are purposeful. The belief that human beings are different from the sum of their parts is given credence when we acknowledge that we are always changing and indeed are, as Rogers affirmed, integral with the rest of the cosmos.

Phillips notes two different perspectives of human beings, which he calls body image and human field image.[38] A body image perspective is the traditional mechanistic way of focusing on dysfunctional parts of the body, while, conversely, a human field image perspective focuses on the whole human potential. Human field images rely on one's never-ending potentials, the unfolding possibilities in the striving toward actualization. Body images rely on the integrity of the physical body and all of its parts in order for health to be attained.

Phillips' distinction between a body image and a human field image view of human beings elucidates the distinction between various notions of how health and healing occur. Thus, we can identify a relation between these two perspectives and the two different ideas of energy. The traditional Newtonian view is equivalent to the body image perspective, which focuses on being. Changes in health and healing are seen as causally related, and energy is exchanged or transmitted between parts. The human field image perspective focuses on becoming. Change emerges from a mutual process where all things are inseparable. In this view, energy is a phenomenon; it is not transferred but rather is transformed.

Of course, I advocate the second view. I used to think that if we could bottle and give "energy" to our patients, they would surely be the better for it. Indeed, the health science literature intimates this could be so. Yet it appears that we have been wrong, that each of us already has energy, because we *are* energy; it is simply a question of whether we are aware of it. It is not a matter of increasing, gaining, or conserving energy; rather it is a matter of changing, of transforming ourselves in ways that will be conducive to well-being.

Perhaps we can't change all things, all the time, because, remember, we are not alone in this transforming process; it is a communal process. Everything changes together and nothing is standing still. What we can do is to assist each other to see more clearly how we can actualize our unique potentials. That's where, I think, Barrett's theory of power as knowing participation in change comes into play. As noted by physician Jeffrey Gold, who practices medicine from a unitary perspective: "We must promote a view of human wholeness in the delivery of health care . . . so that the patient has the freedom and opportunity to make choices and participate in change."[39]

Recipients of health care today need to know that we all participate in this universal process of health and healing. It is not enough simply to do what their health care practitioner says without question; recipients of care must knowingly participate in change. Individuals seeking health care must learn to ask questions and make sense of things, so that each person can choose which actions to take in order to actualize his or her unique potentials. In addition, it seems that we frequently know, without being consciously being aware, what is needed to transform illness into wellness. Perhaps we need to listen and pay closer attention to the subtle signs about us and in us, which often indicate the direction our energy transformation should take. Just as synchronicities appear to have some underlying purpose or meaning, our movements are purposive without our always being consciously aware that we are indeed doing what is needed to enhance well-being.

Physician Howard Cutler quotes Dr. Martin Luther King: "What does not destroy me, makes me stronger." Cutler goes on to note that, "while it is natural to recoil from suffering, suffering can also challenge us and at times even bring out the best in us."[40] So, maybe we are making a mistake to perceive illness as a negative event, for it would seem that positive transformations often emerge from our seemingly negative experiences.

Practitioners need to acknowledge that recipients of care actively participate, perhaps both knowingly and unknowingly, in their own becoming, in the activities related to health and healing. Physicians, nurses, and other health care professionals cannot *effect* change. We can only participate in the inherent, communal process of energy transformation, and sometimes we just have to have faith that it will work out the way it's supposed to. We can't base all of our actions on outcomes, either, because outcomes only indicate a passing moment in an eternal movement. Nothing ceases. Lastly, if we recognize the purposive nature of health and healing for what it is — a striving to actualize unique potentials — we can further assist those who seek to do just that.

But the question remains, Why are we put together so well? Why do we naturally actualize some potentials while others lie dormant? Why all this purposive becoming, human and otherwise? If everything is a manifestation of one thing, and it's all changing together, why isn't there absolute chaos? Why the orderliness? The big *why* question leads us to religion.

Chapter 6

The Spiritual Essence of the Enigma

> God reveals himself everywhere, beneath our groping efforts, *as a universal milieu,* only because he is *the ultimate point* upon which all realities converge.... Our minds are incapable of grasping a reality, our hearts and hands of seizing the essentially desirable in it, without our being compelled *by the very structure of things* to go back to the first source of its perfection. — PIERRE TEILHARD DE CHARDIN[1]

Many noteworthy individuals have tried to bring together the ostensibly disparate ideas of science and religion that exist throughout our world. For my part, it seems very clear that there *is* a point where various scientific views discussed in this book, particularly the newer theories but also some of the old, and *all* religious views converge. If we acknowledge that everything is a manifestation of *one* thing, energy, not only does the indivisible wholeness of it all become evident, but it also becomes, for most of us, impossible to deny the teleological propensity of everything. *Unity and purpose* are everywhere evident, and whether one is enlightened through science and reasoning, mystical awareness, or some kind of religious revelation, all of these paths seem to lead to a divine designer.

The words "unity" and "purpose" aptly express the two fundamental principles underlying religion. Science writer Fritjof Capra and Benedictine monk David Steindl-Rast make a clear case for a universal idea of Religion, emphasizing that a *sense of belonging* underlies all religious views.[2] In this sense Religion is a universal idea that encompasses all the little religions (with a lower-case *r*), that is, the distinctly different religions that exist throughout our world.

Three of the best-known world religions, Christianity, Judaism, and Islam, along with lesser-known religions such as African religious traditions and Native American beliefs, share common ground. They all emphasize community—oneness—and the belief that there is some inherent purpose in the world. In the Eastern religions, particularly Hinduism, there is also a strong sense of belonging, of *oneness*. Even though the Hindus have names for various divinities, their scriptures allude to the idea that there is ultimately one divine reality.[3] In the same manner, Taoism, a Chinese philosophy, advocates that everything is inseparable and that each thing has some underlying pattern which helps to shape and define it in relation with all other things.

Philosopher Renee Weber, in an effort to show that, "unity lies at the heart of our world and that it can be discovered and experienced by man," met with many noteworthy individuals to discuss what lies at the heart of their beliefs and theories.[4] In her book, *Dialogues with Scientists and Sages,* she identified ideas of unity in both science and mysticism. His Holiness the Dalai Lama, when discussing *individual* things and the Buddhist concept of *sunya* with Weber, said:

> They are not self-sufficient but are interrelated. This lack of independence is *sunya*. Emptiness is the absence of independence. In the Buddhist viewpoint, because the ultimate nature is sunya, wholeness comes into being.[5]

Look where you will, there is dispersed throughout all seemingly different cultures, disciplines, and religions, a pervading idea that underlying our reality all is essentially one.

I recently had the good fortune to attend a conference on science and spirituality held at the University of California at Berkeley. The conference was the culmination of three years of work during which prominent scientists and religious leaders came together to examine the relation between science and spirituality at the Center for Theology and the Natural Sciences in Berkeley, sponsored by the Templeton Foundation. The project was guided by the vision of philanthropist Sir John Templeton to promote a link between the sciences and religion, and the subsequent conference was well attended.

Scientists and professionals of every faith and discipline were there and quite respectfully listened to the speakers while they voiced opinions on the mixing and congruency of scientific practice with religion. But one speaker stands out most prominently in my mind, the Christian biochemist Pauline Rudd. Of all the speeches, hers was the one that elicited the most ardent applause, and it was because, like none before or after her, she made a case for unity.

Rudd set out, as she noted,

> ...to explore the idea that the same spirit inspires each of us whether we are artists, scientists or philosophers and that the insights we derive from our various perspectives are all essential if we are to understand even a little of the nature of God and what it means to be human. For in the same way that the body is one though it has many parts, so we are all members of the one body of humanity and the same spirit enlightens us all.[6]

In a similar vein philosopher Alfred Whitehead had written that "the great social ideal for religion is that it should be the common basis for the unity of civilization."[7] How is it, then, that we have forgotten this truth, this human value of oneness? The history of humankind verifies that we have, on numerous occasions, either forgotten or not quite understood that all is essentially one. The Holocaust, the Inquisition, bondage, slavery, and many other horrible atrocities attest to that.

Discussing creativity and the historical function of religion, David Bohm noted:

> A key function of religion was to teach a kind of self-knowledge, aimed at helping man to be whole and harmonious in every phase of life. To this end it was necessary, of course, to cease to be concerned excessively with narrow interests, of self, family, tribe, nation, which latter tend to break the psyche of man into conflicting fragments, making a wholehearted *total* approach to life impossible.[8]

This, I think, has been the problem: the fragmentation of human values has led to apparent disharmony among people. It is not that we can't have different religions or different scientific viewpoints. Rather, it seems to me that we can have different views and yet recognize there is some coherence among all of them. If so, perhaps we can begin to regain the harmony of the whole. Perchance we can agree to say that no one way is right or wrong; rather *every* way is in some way right. The following, then, is no lame attempt to unify humanity. *We are already* a *union.* We are but a small part of a significantly large whole.

Back to the Beginning

> A wheel was shown to me, wonderful to behold.... Divinity is in its omniscience and omnipotence like a wheel, a circle, a whole, that can neither be understood, nor divided, nor begun nor ended.
>
> — Hildegard of Bingen[9]

Recall that many of the ancients were convinced that all natural things are made of a single substance, and Parmenides noted that nothing can come from nothing. If we apply their conviction to the idea that energy is a phenomenon, we can agree that an actuality first had to exist in order for any potentialities to be actualized. And, although the ancient Greeks did not know what this single substance was, they were fairly correct in their belief. Everything that exists is a manifestation of energy.

We have no logical reason to doubt that energy has always existed. It certainly appears that it never had a beginning. On the other hand, the universe arose from some primordial form of energy; the universe did have a beginning in the sense that it has evolved. Nonetheless, one could say that the universe didn't truly have a beginning, inasmuch as the universe was and still is the phenomenon energy; however, it has now evolved or actualized into numerous forms of energy which didn't exist before.

When we think of something arising from nothing, we are picturing substances or material things coming into being from some void, some "nothing." "Nothing" could bring one to think about some "thing" that *is not.* And surely to picture some physical or material "thing" coming into being from some "thing" that is *not* is very difficult. Moreover, some philosophers have argued that if one thinks of some thing that is not, then that some thing *is.*[10] Nonetheless, if we replace the nothing with an intangible something, an imperceptible form of energy, it is easier to think that something can indeed spring from "nothing" or at least "nothing" that is tangible or visible to us. Such as the elusive neutrino that transforms itself into another kind. It's for the most part an intangible form of energy; it can pass through trillions of miles of lead without anyone ever knowing it had passed by.[11]

After the pre-Socratics; Aristotle argued that there must be an "unmoved mover," which is the original, unchanging, source, giving all else in existence the ability to change. The unmoved mover is an *eternal* actuality, an *energeia,* although it is not physical in the sense that all other actualities are. For Aristotle, change in the universe was constant and all changes emerged from the eternal first agent of change.[12] Just as Parmenides had argued that nothing can come from nothing, Aristotle believed that actuality (*energeia*) preceded potentiality. " In other words, we must have an actuality from which all potentials arise.

Aquinas and Aristotle

Aristotle's theory that actualities must precede potentialities was a source of concern for the theologian Thomas Aquinas because he believed that God had created the world out of nothing.[13] No material substance existed before God made it, so something had to have come from nothing. Thus, Aquinas believed that the possibility of creation according to the scriptures was ruled out with Aristotle's conception of change.

If we replace the matter with an intangible energy form, then we still have something actual, even though it may be perceived as "nothing." Now, this intangible, actual thing that appears to be nothing has the potential to actualize, or to become some material substance. Using this analogy, Aristotle's and Aquinas's conceptions of change become congruent. Energy, as actuality, preceded potential matter, or substance. Yet, energy in the primordial form was not a real entity, in the sense of a perceptible thing or substance. It was "nothing," relatively speaking.

We do not know that energy (as an actuality) always existed; God could have created the phenomenon energy from nothing. Energy may have been a potentiality that God chose to create, to actualize. However, it seems much more plausible that the energy was there all along, for if it wasn't we would have to say that the laws of nature were created along with energy itself.[14] And, if the phenomenon of energy wasn't present, then the conservation of energy principle (a law of nature) would have been violated, for something would in fact have come from nothing.

Now *if* the primordial energy was created, then we would have to say that the laws of nature were created as well. And, although I see no reason why we can't assume that God created both the primordial energy and the (natural) laws of nature, some individuals can still argue that there is no reason to suppose that the primordial energy was ever created — that it was there all along — and thus the laws of nature must necessarily be eternal as well. Indeed, Linde did suggest that the universe is immortal, and Prigogine remarked that, "although our universe has an age, the medium that produced our universe has none. Time has no beginning and probably no end."[15] Either way, it seems clear that something must have been at the helm; the self-organizing ability of energy is much too neat. The fact remains that everything we know to exist arose from some *one* phenomenon, some one thing. And it has evolved into millions, even billions, of other things which evidently didn't exist before (if there was a "before").

Genesis, Time, and Energy Transformation

It is said in the book of Genesis that God created us in his own image. Bearing this idea in mind one could also argue from the Aristotelian thesis as follows. God, the eternal unmoved mover, is pure *energeia* (actuality); we are *energeiai* as well. We *were* created in God's own image. God is the ultimate alchemist and has created us, alchemists as well, in the same image. Hence, we are led to say perhaps that God is energy. As energy God has existed eternally and, in a sense, we have existed eternally, for we are all of one with God. Phillips likewise notes that "God is... energy, along with the human and environmental energy fields."[16] Yet it seems clear that the ultimate alchemist is above and beyond that which has been created.

There has always been an obvious dissonance between theism, pantheism, and panentheism. To me it appears that God is immanent in *all* things but also beyond *all* things; God is transcendent. For it seems that nothing less could have designed this marvel of a universe with all of its enigmatic features. Nothing less could give purpose to the smallest, or inanimate, or seemingly useless things. Yet the idea of energy as a phenomenon does not contradict other religions which claim that God is identical with everything in the universe (pantheism), or the belief that there is a universal sentience which pervades all things (panpsychism). Nor does it at all eradicate the panentheistic belief that God is in all things. What we can intuit from the idea that energy is a phenomenon is a convergence of different viewpoints which may ultimately lead us to a better understanding of the whole. So I leave it to go where your heart leads.

Many scholars today commonly accept that the book of Genesis need not, nor can it, be taken literally. For example, science has supposedly disproved the possibility that God created the universe in seven days. However, it is difficult to understand this because we really use time as a convenience. After all, Einstein demonstrated that time is relative. The light of stars takes many billions of years to travel to earth, so when we look at a star we are seeing something that existed eons ago; yet, in our existence, the star whose light we now perceive still exists (in our reality). The stars are in the sky as far as one can tell, yet scientific study has led to the discovery that many of the stars we see no longer exist in *our* here and now.

There are those who maintain that the book of Genesis is a fable because it says that God created the universe in seven days. Yet most of these same people assert that time is not absolute. So what if, as some believe, there are multiple realities and time is relatively different for each of these realities?[17] What, then, might this imply? It has been established that clocks in space run differently from those on earth. What, I wonder, is one minute, one day, or even one year of our time, equivalent to in God's time? If time is no longer held as absolute, why is it not possible that God created the universe in seven days according to God's time? In fact, St. Augustine said that God created everything, including time, simultaneously.

Again I am reminded of the mystics and their notion that time is an illusion. Meister Eckhart wrote:

> Perception means here seeing in the light that is in time, for anything I think of I think of in the light that is in time and

> temporal. But angels perceive in the light that is beyond time and eternal. They know in the eternal now. . . . Yet take away this now of time and thou art everywhere and hast the whole of time.[18]

Time itself is an enigma. There have been many books written concerning the nature of time, and although the subject takes us a bit off the course, perhaps another small divergence is warranted here.[19] Paul Davies examined the implications of Einstein's relative time. He pointed out that over the course of history many people have believed that time exists only in the human mind. I found one of his comments particularly insightful: "If time is something measured with clocks, and if clocks *vary* with time, how can we ever know what time it *really* is?"[20]

If science has indeed substantiated that time is not absolute, then the seven-day time period, during which God is said to have created the universe, could really mean seven minutes, seven years, seven million, or even billion, years. Noah was said to live six hundred years and Methuselah a thousand years. How can we claim this is not so, if there is even the remotest possibility that time during their day was different from ours?

Discussing the Kabbalah, the work on which mystical Judaism is based, Rabbi David Cooper notes, "The essential paradox of creation is the issue of how unity becomes multiplicity."[21] I have taken his words a bit out of context since he was discussing religion from a religious view, not necessarily in this instance a scientific one. But his point is well taken. How *does* the One become many? I think science has, in a basic way, solved this dilemma for us.

Going back to the book of Genesis, most scientists also say that we can't continue to fully espouse the story of Adam and Eve, since humans evolved from other living things. Yet it can still be argued that God created what was to become human, if we use another analogy here. In the book of Genesis it is said that man was created from the dirt of the earth. This purportedly contradicts Darwin's theory of evolution. But I fail to see this contradiction. Isn't dirt a form of energy? Aren't all living things essentially composed of energy?

If humans were created from the dirt of the earth, then one form of energy was transformed into another form of energy. And let's not forget that the big bang theory asserts that we evolved from stardust.

This does not contradict Darwin's theory, which is fundamentally saying the same thing. The difference lies in the terminology, not in the general idea of evolution. Humans evolved from simpler species, which evolved from even simpler species. And those simpler species evolved from inanimate matter—the dust of stars and the dirt of the earth. In fact, Teilhard de Chardin, philosopher, scientist, and Christian clergyman, referred to evolution as "transformism."[22] Genesis, the big bang theory, and Darwin's evolutionary theory are all speaking of the same thing: energy transformation. All is essentially one, as Meister Eckhart forcefully stated: "All things are contained in the One, by virtue of the fact that it is one, for all multiplicity is one...."[23]

The Argument over Darwin's Theory of Natural Selection

Darwin's theory of natural selection continues to elicit some debate over whether his theory supports some kind of teleology and ultimately a divine designer. In fact, every few months or so, an article pops up somewhere discussing the continuing battle about whether schools should be teaching children evolutionary theory.[24] And that's partly because many individuals continue to say that Darwin's theory rules out creation according to the Bible. Furthermore, there are those such as Richard Dawkins, professor of zoology at Oxford University, who provide what some consider a plausible explanation for there not to have been a designer, and they base their ideas on the theory of natural selection.

Dawkins bases his argument on the premise that "the essence of life is statistical improbability on a colossal scale."[25] He insists that cumulative natural selection is the only theory that can completely explain life's complex design, and in so doing the theory also results in the abolishment of any deities. Dawkins notes that an organ such as the eye could not have spontaneously developed in its final form for the odds of this occurring are "astronomically long."[26] He concludes this line of thought by saying: "The same applies to the odds against the spontaneous existence of any fully fashioned, perfect and whole beings, including — I see no way of avoiding the conclusion — deities."[27] His argument undermines the obvious purposiveness in nature. Furthermore, when examined within the context that all things are a manifestation of one thing, energy, his conclusion necessarily becomes a blatant violation of the energy conservation principle.

First, Dawkins essentially wipes out any claim that there are teleologic implications in Darwin's theory. He views the theory of natural selection as a sequence of cause-and-effect occurrences, and even though he points out that evolution has not occurred by random processes, Dawkins still insists that there is no purpose underlying the becoming of things. But then it does seem to me that we are actually, in a roundabout way, back to the idea of randomness in nature having resulted in highly complex beings, which Dawkins himself claimed was statistically improbable.

For me the question is how *could* such perfection evolve with no purpose, no reason? And if there is "something" that has created it, doesn't it make sense that this "something" must be immensely intelligent to put things together so beautifully? Second and more to the point, spontaneous existence was never posed as a prerequisite for God. Spontaneity is irrelevant when we speak of a divine designer. God is eternal; God has always been and will always be, but everything else can come to be however God wishes it to do so. We don't have to have spontaneity to conclude that there is some inherent purpose in things, and once we come to this realization it is difficult to say it all came about without something directing it. And finally, if all of this becoming didn't evolve from *something*, it came from nothing, which is a direct violation of the energy conservation principle.

Richard Dawkins' book *The Blind Watchmaker*, in which he discusses his view, is actually a rebuttal of William Paley's famous watch analogy. Using the analogy of finding a watch in a remote place, Paley argued that the human eye, standing alone, without any other examples of highly ordered and complex things, surely must give rise to the idea of an intelligent designer. Likewise, "If there were but one watch in the world, it would not be less certain that it had a maker."[28] Many individuals believe that Paley made a valid point: one cannot discount the beauty, complexity, or purpose, of things relative to our existence.

Philosopher of science Michael Polanyi uses a similar analogy with his example of a stone arrangement at a depot located in Abergele which says, "Welcome to Wales. . . ."[29] He points out that it is obvious that someone intentionally placed this orderly arrangement of stones there because the probability of stones naturally occurring in such a meaningful pattern is "fantastically small." However, Polanyi makes another fine point: had there been no intelligible pattern in the stones, the question of whether the stones were placed there accidentally or not is really of no concern and would, in fact, never even come up. In other words, we must first recognize an implicit or explicit orderliness in things before we decide to examine if its occurrence is accidental or not. According to Polanyi recognition of pattern and order is a reflection of the personal knowledge that we come to acquire through our life experiences.

The foregoing discusses the continuing concerns underlying and provoking the argument over *The Origin of Species.*[30] The debate is closely linked with theology and as such it is doubtful that any one person can sway the argument further in either direction. It is doubtful because religious beliefs are personal. Each of us must come to our own conclusions; such is the uniqueness of humans.

What, Exactly, Is Nature?

The word "nature" has come up often in relation to the enigma of energy and ultimately prompts the question, Just what is *nature*? Historian and philosopher of science R. G. Collingwood notes three periods in the development of the idea of nature, which coincidentally seem to reflect the development of ideas of energy. In his discussion of the first period, the Greek view of nature, Collingwood points out that the ancient Greeks believed a certain vitality or ceaseless motion existed in nature, which they generally attributed to the soul.

Returning to Aristotle's philosophy, I noted in chapter 2 that "*energeia* appears to have been used by Aristotle to express both a tangible and intangible quality that constitutes the very essence of nature." The most important aspect of Aristotle's conception of nature lies in his belief that all things have a *final cause,* which is exhibited by the individual thing's form. According to him the soul was the essence of living things, and of course the *form* of anything was the purpose or reason for its becoming. Overall, according to Aristotle, the teleological qualities of things were so strong that there could be no explanation for anything in nature, including us, without it. Hence, he argued against those such as Empedocles who believed that things come to be by chance:

> Take teeth, for instance: what is wrong with the idea that the front teeth necessarily come through sharp and suitable for biting, and the back teeth flat and good for crushing food? Why should there be purpose behind this? Why should it not just be an accident? And the same question could be asked about any other part of the body which seems to have some purpose. So where every part turned out to be just as it would have been if it had some purpose, the creatures survived because, spontaneously, they happened to be put together in a useful way. But everything else has been destroyed and continues to be destroyed, as Empedocles says of his "cow-like creatures with the heads of men."[31]

In the second period, which Collingwood referred to as the Renaissance view of nature, mechanism was firmly established. This period, comprised of two distinct stages, extended over the sixteenth, seventeenth, and eighteenth centuries, and was also the time during which the term "energy" reappeared in the scientific literature. The first stage was a remnant of the Greek view in which nature was still viewed as "a living organism, whose immanent energies and forces are vital and psychical in character."[32] Although this stage was similar to the Greek view, in certain respects it differed. There was a rejection of final causes, and instead the emphasis was placed on efficient causes. However, because nature was still viewed as an organism, there was a mysterious aspect to it as well, which gave rise to the idea that magic and astrology played a part in humankind's relation with nature.

Collingwood notes that the second stage of the Renaissance view of nature came about with the Copernican discovery that our world was not the center of the universe. The main contention during this time became "the denial that the world of nature, the world studied by physical science, is an organism and the assertion that it is both devoid of intelligence and of life."[33] It was believed that nature arose from God's intervention. God created humankind, and humankind in turn created machines. It was asserted that humans could control aspects of nature. This may also have had a connection with the passage in the book of Genesis, which has often been erroneously interpreted as God giving man "control" over nature. But there are many ways to interpret this passage.[34]

During this period, human beings were seen as outside of, rather than a part of, nature. We became pompous, thinking that we controlled things and that we were somehow superior. Explicit in this view was the denial of final causation. The primary focus was on matter and the natural laws by which matter changes. Science and philosophy recognized only efficient causes: forces producing effects. And, finally, mathematical structure accounted for the changes, both of a qualitative and quantitative nature.

The third and last period identified by Collingwood, the modern view of nature, had its origin in the latter part of the eighteenth century when process and change rather than the elements (which themselves did or did not change), became the focus. This period, not surprisingly, coincides with the emancipation of the energy conservation doctrine. I believe that during this period the idea that energy was an autonomous existent contributed to the shift in focus. It became vaguely evident that change was inherent in various things; that is, it was recognized that change could occur without the provocation of external forces or efficient causes.

Collingwood identifies the idea of a "rhythmical pattern" with the modern view of nature and acknowledges that the new physics theories are partly responsible for this notion.[35] But the rhythmical patterns we now know to exist in nature also seem to denote an inner principle of change, or an Aristotelian "that for the sake of which," originally expressed by the ancient Greeks. So one might say that we have come full circle.

In conjunction with this new take on an old idea that was present in both Eastern and Western antiquity is the increasing awareness that intuition plays a significant role in scientific discoveries. As the historical background of the idea of energy attests, intuitive ways of knowing have been crucial to the development of scientific ideas throughout history. Many individuals knew things, such as the energy conservation doctrine, without being able to empirically verify them. In other words, intuitive ways of knowing seem to have led us in the right direction long before we were capable of scientifically validating what we somehow knew to be so.

Subjectivity and subjective ways of knowing, such as intuition, have become as vital to our understanding as objectivity and empirical ways of knowing. In this modern view of nature humankind has once again come to be recognized as being a part of nature, rather than outside of it. As Prigogine notes: "Nature is part of us, as we are part of it. We can recognize ourselves in the description we give to it."[36]

We may be at the beginning of a new cosmological period, although I hesitate to name it, for I'm still not sure exactly what the new epoch entails. But it is obvious that when we speak of "nature" we are actually speaking of the multiple manifestations of energy and its ongoing purposive transformation. This eternal energy transformation is a collective process rather than an exclusive, or restrictive, cause-and-effect, one.

The whole question of nature is clearly related to Capra and Steindl-Rast's idea of *belonging to the universe,* and in turn it ultimately leads us back to the topic of God. Returning to the passage in Genesis, which seems to imply that humans have control over nature, Capra and Steindl-Rast note:

> We are put into the Garden "to tend to it and keep it." That's where the sense of responsibility comes in. Stewardship, rather than exploitative domination, is the crux of that passage. Unfortunately, the Book of Genesis has been misinterpreted and misused to justify a lot of destructive things. Rightly understood, Genesis says that Adam is put into the Garden to tend it and take care of it. So that is our responsibility.[37]

It is clear that many people have misinterpreted the text; our suffering oceans, rivers, rain forests, and other natural environments can attest to that. However, even with reverence for nature, since we are part of nature, it may be quite difficult to know when we are properly taking care of nature. That is, *how* do we *know* we are doing the right thing?

I am reminded of Lorenz's statement that the butterfly flapping its wings in one place may lead to foul weather in another place somewhere in the future. And I recall the well-documented evidence that a human body full of cancerous cells can develop an infection, and then shortly thereafter the terminal illness subsides or disappears. How do we ultimately know we are doing the right thing, ever?

When my youngest daughter was walking home from school one day, she encountered a baby bird on the sidewalk. She came home very upset about this baby bird and cajoled her big sister into driving her back by the school to get it. They brought it home and proceeded to call our vet, go to the pet store, and literally set up house for this baby bird. The bird survived for only one day. We couldn't get it to eat anything. It chirped and drank water from a dropper, but it finally died.

Someone once told me that mother birds will occasionally "throw" some of their newborn offspring out of the nest, that they sometimes choose certain babies to discard. How do we know if this mother bird somehow knew this particular baby of hers would not survive? Perhaps in the general scheme of things, it just wasn't meant to live. We humans cannot ultimately know what's right or wrong when we intervene in nature. That's where responsibility comes in. It seemed the right thing at that time to help the baby bird.

While I was on the phone with a friend the other night, my teenage daughter was heard yelling in the background. When I was asked why she was yelling, I told my friend about the two slugs that were out on our front porch. Every summer for as long as we've lived here, the slugs visit our porch. My daughters always yell when they encounter any kind of crawling thing. I have been known to yell for the same reason. My friend proceeded to tell me to put down salt and it would get rid of the slugs. I could only wonder why I would want to do that.

The Christian mystic Angelus Silesius wrote:

> Nothing is imperfect, the pebble equals the ruby
> A frog is as beautiful as any seraphim.
> In God all is God; the simplest little worm
> Is as much in God as thousands of Gods.[38]

For a long time I have believed that all things have some purpose; if they didn't they wouldn't exist. I don't like the slimy looking things any more than anyone else; yet they don't cause me and mine any harm, they aren't depreciating my property, they aren't *doing* anything to us. Why should I do anything to them?

We have a responsibility to do what we consider to be right, but, on the other hand, unfortunately we may never really know just what *is* right.[39] Ultimately we have to have faith in the purposive manner in which the innumerable forms of energy actualize certain potentialities while limiting others. And faith leads us back to religion. Steindl-Rast notes that "faith . . . is a matter of trust. Courageous trust in that ultimate belonging which you experience in your religious moments."[40] Yes, that is how I see it. When we trust the natural manner in which the phenomenon of energy transforms itself all around us (and in us) into meaningful patterns, we experience a revelation that *God is* and *is with us.* Said another way, God is revealed in the innumerable, purposive manifestations of energy. My faith arises from this revelation; I can see it no other way.

Human Nature and Nature: All Essentially One

In the most basic sense science is the study of nature and the human health sciences are the study of *human* nature. When we speak of *human* nature we are referring to the inherent tendency of humans to be the way we are, to do the things we do. Just as the *nature* of humans is likened to purposeful actualities, so too does the idea of *nature* in the larger cosmos intimate purpose. As I mentioned earlier, there is evidence of changes in human nature, which some, like nurse theorist Rogers, describe as an increasing diversity in human life forms.[41]

It may be that Bergson and Teilhard de Chardin were right. We may be on the edge of a new reality where we are increasingly transcending the confines of physicality. We may all one day be mystics, in a perpetual state of what physician Richard Bucke referred to as "cosmic consciousness," where we will *all know* irrevocably that all is one, that it is eternal and:

> . . . that the universe is so built and ordered that without any peradventure all things work together for the good of each and all; that the foundation principle of the world is what we call love, and that the happiness of every individual is in the long run absolutely certain.[42]

Likewise, Watson linked human nature and our potential to actualize various paranormal abilities with the growing awareness of the unity in nature. He writes: "Or greatest strength lies in unity . . . here on earth, and the other is that this unity could give us the impetus we need to transcend the system all together."[43]

After all this is said, one thing still remains. We are all manifestations of energy, and hence it is all essentially one. No one has expressed the idea of oneness quite so clearly as Father David Toolan:

> And what are we humans, the observers, in this perspective? We come at the end of the cosmic meal, at the end of an enormously complex chain of conversions of noisy energy radiating to us in torrents and stretching back to the formation of galaxies out of the energy released by the big bang. Quite literally, we are the fallout of stars, for without star factories converting helium out of hydrogen, there would be no oxygen, carbon, or iron; and without them there would be no amino acids or proteins for life. . . . Like the stars, we, too, are dynamic energy systems, internally constituted by our relationships. We are open . . . systems moving upstream and drifting towards death . . . like the Sun and the Moon, we are disturbances in the field, vortices in turbulent nature. Like them, we are constantly recycled, dissolving and reappearing. We regrow our entire physical body as we do hair and nails. Nothing in our genes was present a year ago. The tissue of our stomach renews itself weekly, the skin is shed monthly, and the liver regenerated every six weeks. Every moment, a portion of the body's 10^{28} atoms is returning to the world outside, and ninety-eight percent of them are replaced annually. Each time we breath, we take in a quadrillion atoms breathed by the rest of humanity within the last two weeks and more than a million atoms breathed personally by each person on Earth. So much for the strictly bounded, separate individual.[44]

Yet there is a continual eternal movement in the evolutionary process, so perhaps we shouldn't think we come at the *end* of the cosmic meal. And there also remains death — the meaning of which is generally assumed to be the cessation of life. If our constitution, the matter of our being, recycles, dissolves and then reappears, how can the ultimate dissolution of our bodies indicate the absolute cessation of *life?* Or maybe we need a new definition of life (or of death). Perhaps we need to include death in the life process, since dying appears to be a means of further actualizing our potentials.

Death and the Eternal Soul

Dying as an innate aspect of living is not easily accepted in many societies, particularly in this modern age where health care professionals often act as if death is the enemy. As technology has advanced, we have progressed to the point where we can honestly, or so we believe, prevent patients from dying and in many situations bring patients back to life. So that dying, for many, has become a cold, isolated process—cold in the sense that it is hard-core machinery touching the patient rather than the gentle grasp of a familiar loving hand; isolated in the sense that strangers rather than family and friends often surround patients in their last living moments, strangers who are valiantly trying to prolong life. Where has the humanness gone?

Truth be told, although many individuals may not believe that the essence of their being survives physical death, they will still acknowledge that dying is part of living. It is the rare person who would consciously choose to die among strangers, and yet in many areas of health care today we, in essence, force those who are dying to do so. What is called for here is that we acknowledge the essential oneness of it all, while retaining respect for the unique needs of the individual. Everyone needs to feel a sense of belonging, in living and in dying.

It would be remiss of me not to at least mention the idea of immortality here, even though I discussed the idea of the eternal soul earlier. I have already alluded to my belief in everlasting life. I became utterly convinced with my experience of the psalm. No matter how I rationalize it, seeing a truck with Psalm 122 posted on it at a time when I was desperately missing my "dead" father, was, at the very least, a sign that there is more to living and dying than what we currently know. I simply cannot believe that the timely appearance of the psalm was coincidence.

In the Kabbalistic tradition they write of a soul level called *yehidah*, generally meaning unity. Rabbi Cooper writes:

> *Yehidah* is our ultimate link with God-ing, the part of us that can never be separated from the Divine. When all else fails, the awareness of our *yehidah* endowment may be sufficient to carry us through our more difficult times, for it is the aspect of each person eternally connected with perfection.[45]

Our souls, our essence, are always part of the One. Bergson wrote in a similar fashion about the mystical experience and the uniting of the soul with God. We are intricately *eternally* connected; it is all essentially one, and as such nothing truly ever dies.

Many scientists and philosophers believe that consciousness survives physical death. I expect that continuing research in the area of parapsychology and near-death experiences will somewhere down the line add weight to the idea of the nonlocal mind and thus the eternal soul. A final issue is whether or not the eternal soul is applicable only to human beings. Of this idea Thomas Matus, a Benedectine monk said:

> That's the old paradigm, that your pets will not be in heaven. That's one of the most terrible things ever said to children! This is not theology. This is cultural baggage, a whole collection of trinkets that is not theology.[46]

Steindl-Rast adds that resurrection, and consequently salvation, is a kind of "cosmic renewal."[47] That, for me, is the essence of the eternal soul. All of nature, and humans are part of nature, is constantly being renewed; it is manifested by the continual transformation of energy.

Summary

Reasoning, supposedly a uniquely human ability, leads us to question the "how and why" of existence, particularly human existence. And orderliness, of the magnitude that we observe all around us, has lead many of us to reason that there is some inherent purpose in nature. In turn, at least for many individuals, this teleology leads to God, as reflected in a comment from Henslow to Darwin in 1831:

> Can we wonder, then, that Nature's productions should be far "truer" in character than man's productions; that they

> should be infinitely better adapted to the most complex conditions of life, and should plainly bear the stamp of far higher workmanship?[48]

The way I see it, the argument of whether religion has any factual basis consists of the following. Either existence, and by existence I mean everything there is, is purposeful, or it is not. Many have said that evolution is, by its very nature, teleological. If certain forms of energy cease to exist while others prosper and grow, does that not imply purposiveness? If change occurs, and that is precisely what evolution is all about, how can one say it does not occur for some reason or, rather, some purpose? Simply put, would there be such a thing as "change" or, for that matter, progress (equivalent to positive change), if there were no reason for the occurrence?

A growing number of scientists and philosophers are beginning to view, or at least allude to viewing, our universe as orderly and, thus, purposeful. The physical theories that grow by the day seem to attest to some underlying meaning in it all. It, the becoming (our becoming) continues to be a simultaneous communal process of evolution. Once we acknowledge that all of nature is essentially one, there can be no doubt that we have a responsibility to actively participate *with* the rest of nature: to actualize the inherent potentials throughout our world. Father Toolan asks: "Do you not hear the stars, quarks, rain forests, and whooping cranes calling out to you to name them properly, to give them voice, to put heart into them by what you make of them?"[49] I do. We are all in this becoming, together.

While many of us have used our reasoning abilities and intellect to slowly make our way to the realization that all is essentially one and arises from God, the mystics throughout time have always been acutely aware of it. Evelyn Underhill, a great mystic of the twentieth century, examined the interrelation between mysticism and theology. She noted that mystics have come from all religious backgrounds so that we could never "limit mystical truth — the direct apprehension of the Divine Substance" to any one religion. Yet, she notes, "the gold from which this diverse coinage is struck is always the same precious metal: always the same Beatific Vision of a Goodness, Truth, and Beauty which is *one*."[50]

In concluding my speculative comments on significant theological and teleological implications relative to the enigma of energy, I am reminded of a comment made by Peter Berger in *A Rumor of Angels:* "Most of what we 'know' we have taken on the authority of others, and it is only as others continue to confirm this 'knowledge' that it continues to be plausible to us."[51] Maybe, as Berger implied, we are socialized into believing or disbelieving various things in life. Surely we cannot dispute that the socialization process and, indeed, historicity, do help to illuminate the nature of things. However, we are without question highly intricate beings with an inordinate capability for *reasoning* and for *questioning*. This makes it difficult to believe that we have been socialized by the course of history to accept or believe that orderliness in nature exists, since, if we look, we cannot fail to see the signs all around us.

Can we profess that there is no element of teleology to all this? Certainly if form arose from formlessness, it implies that the purpose was there all along. From one spark of energy, one form of energy . . . came all these other manifestations of energy. The various sciences continue to search for, and empirically demonstrate, that order does indeed exist on all levels in our universe.

In *Fire in the Mind: Science, Faith and the Search for Order* George Johnson writes, "It is hard not to believe that we are all participating in something universal, something holy, that the pageant must unfold beyond our planet."[52] As Hawking notes: "The whole history of science has been the gradual realization that events do not happen in an arbitrary manner, but that they reflect a certain underlying order, which may or may not be divinely related."[53] Therefore, it seems it is *reasonable* to say, "Purposiveness exists." Now we just have to reason "why," and that, I maintain with Hawking and Davies, is in the mind of God.[54]

Chapter 7

Full Circle: Back to the Enigma

> The most beautiful experience we can have is the mysterious. It is the fundamental emotion which stands at the cradle of true art and true science. Whoever does not know it and can no longer wonder, no longer marvel, is as good as dead, and his eyes are dimmed. . . . I am satisfied with the mystery of the eternity of life and with the awareness and a glimpse of the marvelous structure of the existing world, together with the devoted striving to comprehend a portion, be it ever so tiny, of the Reason that manifests itself in nature.
>
> — ALBERT EINSTEIN[1]

We have examined many areas relative to the concept of energy from antiquity through modernity, so, some may ask, exactly what *is* the enigma and is it still there? The answer to these two questions, I think, will depend to a certain extent on the way energy is viewed. For me, the answer is unequivocally, yes; of course the enigma is still there. If asked what it is, I would have to say that the mystery underlying the phenomenon of energy is the apparent oneness of it all and at the same time the distinctly unique things it has become and, finally, the purposive manner in which energy actualizes its potentials. More importantly, I would have to say that ultimately the mystery must lead to something far beyond what we know exists, beyond this universe, this earth, and us. I choose to call this something God.

So we have come full circle. The various scientific and philosophic views, when combined, have in a sense revealed to us what the enigma of energy is, only for us to find that a deeper, far more enigmatic entity underlies our dilemma. For no one can render ultimately intelligible what God is or why, for that matter, we and everything around us exist. Yet it is clear that both science and religion have gone along similar paths, which may or may not be converging, depending upon how you look at it. To me it seems that they have already converged. Teilhard notes: "Science (which means all forms of human activity) and religion have always been for me one and the same thing; both have been . . . the pursuit of one and the same Object."[2] With that in mind, let us turn once more to the enigma of energy.

What Is Energy?

The analysis and interpretation of the literature undertaken here point to the necessary task of reconceptualizing the idea of energy to more suitably fit modern ideas of nature. Although the basic idea of energy as a thing (which manifests itself in a multiplicity of forms) has been evident to some degree since the ancients, the use of the word and hence the concept has been inconsistent and has lacked clarity. In fact, many dictionaries today, such as *The Oxford Dictionary and Thesaurus,* still equate "energy" with "force."

We have two distinct ways to view and conceptualize energy: as part of a causal process and as a thing, a phenomenon. The ideas of energy as part of a process and as a phenomenon both appear to have originated in Aristotle's conceptualization of *energeia* and have remained vaguely present throughout history. These two different ways of conceptualizing energy are not congruent.

The first and more strongly emphasized meaning of *energeia* in Aristotle's philosophy is that of any physical actuality having a distinct form. Therefore, for modern day conceptions of energy to be congruent with Aristotle's *energeia,* things must be *physical* entities. Now, today we know that some manifestations of energy do not appear to be physical, or at least tangible, an obvious example being the various fields discussed previously. This idea is concisely expressed by Zukav: "electromagnetic fields involve no object whatever"; yet, these fields are real entities.[3]

The electromagnetic and morphogenic fields, are basically intangible forms of energy, but they are actualities nevertheless — just as wave-particles can be tangible or intangible, just as matter is tangible and the hypothetical ether isn't. The basic idea that *energeia* is a *real thing* is quite explicit in Aristotle's philosophy and it has been carried over into the modern-day concept of energy. The idea that energy is a phenomenon, a *thing,* is implicit and, at times, explicit in various scientific disciplines.

The second way Aristotle used the term *energeia* was as part of the process of change. When Aristotle spoke of *energeia* as process, he broadened the scope to include potentialities; that is, he spoke of things actualizing their potentials. Here one might see an allusion to the transformational ability of *energeiai,* in that, by actualizing potentialities, things change.

Recall Aristotle's idea that change (*kinesis*) is the actuality (*energeia*) of that which exists potentially, which leads one to believe that all change and motion is *energeia.* So, as Ackrill notes, "*Kinesis* is itself an *energeia.*"[4] In modern literature, energy is often referred to as the ability to perform activity or work. Change comes about through energy usage.

We can use the example of hydrogen and oxygen to see the difference between these two views of energy. In the idea of energy as part of a process, two hydrogen atoms and one oxygen atom combine to form water; the combining is an energy process. This view of energy is mechanistic in that it is believed that one separate thing acts upon another to cause change; energy is used accordingly. In the idea of energy as a phenomenon, we could say that the hydrogen and oxygen are forms of energy, and in concert they may combine (and often do), forming water (or ice or steam), yet another form of energy. The changing nature of these manifestations of energy is a characteristic of the hydrogen, oxygen, and, obviously, the water. This view of energy is dynamic; the changing nature is within the phenomenon of energy rather than outside of it.

How and Why Things Change

In his search for the reason why *energeiai* — actualities — tend to become certain things and not others, Aristotle derived from both reasoning and observation that there must be a reason why things ever change at all. Consequently, he developed the doctrine of four causes: (1) the material cause is the matter without form, for example, a piece of marble; (2) the formal cause is the pattern or form that the marble will become; (3) the efficient cause is the hammer, the chisel, and the sculptor's use of these tools; and (4) the final cause is the purpose or reason for which the change takes place, the "that for the sake of which," in this case, a statue.

The *form* is implicit in the matter, or perhaps it is better to say that in the matter, the marble, lies the potential for the form, the statue. It is only when matter is combined with form that one may call it a complete actuality (*energeia*). However, in the actualizing of the potential matter into the form of the statue, we have a process of change, which can be equated with Aristotle's looser definition of *energeia*.

In many of the scientific theories that came after Aristotle's time, the efficient cause was emphasized, and the "that for the sake of which" became an antiquated idea. Yet, it does seem as if Aristotle was onto something. He presupposed that all things in nature have an underlying pattern or orderliness and that all things appear to progress, to actualize potentials, in a meaningful manner. Thus, all things have an underlying purpose, a distinct pattern of direction.

Decoding the Pattern

In chapter 2 I quoted Morris West, who said that "there is a pattern in all this which I begin slowly to understand and which I am trying to codify. The problem is that I lack the vocabulary."[5] It seems that there *is* a pattern in all this, and we have not lacked the vocabulary to describe it; rather, we have *disguised* the pattern. We have hidden it by fragmenting ideas, by partitioning disciplines, cultures, and religions.

In a discussion of causality and synchronicity David Peat refers to variation principles, scientific principles of wholism which predate Newtonian science. Without getting into the complex nature of these principles perhaps we can still see a relationship between them, Aristotle's philosophy, and the idea of energy as a phenomenon. Peat notes that, generally speaking, variation principles describe motion or activity, but they do so "by recourse to their whole action or gesture."[6] They show, in Peat's words, "that all motion and change emerges out of a 'law of the whole' and that the patterns and events of nature are the expression of an underlying unity of form."[7]

There is an underlying unity of form behind all motion and change. Just as Aristotle saw a "that for the sake of which" in every "thing," so too can today's physics be said to imply that there is some inner principle of change underlying the becoming of it all. For the becoming of it all, the becoming of every *thing*, has emerged from one continually, eternally changing *whole.*

Many of the scientific theories discussed previously indicate oneness and purpose in nature, for example, Bohm's implicate and explicate orders, where everything arises from one common ground. Other examples that we have reviewed include chaos and complexity theories, where there are indications of self-organization and underlying order, as well as Sheldrake's theory of morphogenic fields and the newer quantum theories. Briggs and Peat discussed many of these theories, calling them, "looking-glass theories." They note: "Whatever fate the looking-glass theories may suffer in the vicissitudes of science, they certainly bring a conceptual transformation. For thousands of years wholeness has been mute. Now it can speak."[8]

Nonlocality is a particularly striking example of the apparent unity in nature. How *do* split photons stay connected, even despite long distances? Gary Zukav, author of *The Dancing Wu Li Masters,* notes that these events not only indicate that are there no separate *parts* in nature but they also imply that things we see in nature are *not* due to chance. He writes:

> What happens . . . is intimately and immediately connected to what happens elsewhere in the universe, which in turn is intimately and immediately connected to what happens elsewhere in the universe, and so on, simply because the "separate parts" of the universe are not separate parts.[9]

Energy manifests itself in an infinite number of ways, and the forms themselves are always changing. But they are all connected, they are all manifestations of *one* thing and hence act in concert, providing the cosmos, and us, with a harmonious and purposeful existence.

The point at which we are forced to depart from Aristotle's ideas is in his conception of efficient cause. For if one holds the belief that all is one in an essential sense, efficient causes cannot exist. They are an illusion. One thing cannot be *acting* on another thing to *cause* change. Change emerges out of the whole.

The notion of causality has been one of the most dominant ideas in the history of science, and yet there is no substantial evidence for it. Many individuals have argued against causality, among the most notable the philosopher David Hume. He came to the conclusion that causes and effects are merely changes that we find constantly conjoined.[10] Hume professed that "experience only teaches us how one event constantly follows another, without instructing us in the secret connection which binds them together and renders them inseparable."[11]

We can perhaps distinguish two distinct categories of energy forms: self-created and thus self-organized forms such as our universe and all natural inanimate and animate things in it, and those forms that are perceived to be created by human beings. But humans haven't really created them; that is, we can't really create anything alone. It has been and continues to be, a communal, creative process among all the forms of energy, which move in a definite direction to further actualize the potentials throughout the cosmos. The directionality we can attribute to this becoming cannot be causal or linear. Rather, it is a direction that we can perceive arises out of the purposive activity of the innumerable manifestations of energy *together.*

Energy can transform itself from one thing to another, naturally, for example, orange seed which becomes a fruit-bearing tree, or it can be transformed by human means, for instance, oil, which we use to heat boilers and fuel automobiles. Within the energy itself there is an inherent ability to transform, to change as needed. Whatever else it becomes, it becomes for a reason, not necessarily recognized or understood by us humans, who are, after all, another manifestation of this phenomenon.

Energy is the *thing* which has the ability to perform work or be active. It is not *just* the ability to perform work or be active. Energy as a phenome*non* is the substance underlying all that is, was, or ever will be. It exists in a multiplicity of forms and has within itself the ability to change. Its changing nature leads to what are commonly referred to as energy processes, and here scientists have sometimes been able to measure and quantify these processes of change.

So, we humans hold no dominance over nature, for we are a part of nature. Change, the actualizing of potentials, comes about by a concerted communal process. There is no *from* anything; there is only becoming. The implications are profound on a scientific level, but perhaps even more so on an ethical level. For if I hurt someone or something, I am hurting myself; in fact, I am hurting everything that exists. As Bohm said when discussing his idea of the implicate order:

> To see that everybody not merely *depends* on everybody, but actually everybody *is* everybody in a deeper sense. We are the earth, because all our substance comes from the earth and goes back to it. It is a mistake to say that it is an environment just surrounding us, because that would be like the brain regarding the rest of the body as part of its environment.[12]

The ideas of wholeness and of something underlying it all assuredly ring true. So we come now to *living* the talk.

While driving to work one day a few years ago, I looked around at the other drivers and suddenly realized that I must have a great deal of faith in the driving ability of the other individuals driving around me. Yet they are strangers to me. Indeed, all of us who drive must have a great deal of trust in those who drive around us, for they are in essence driving *with* us. We trust that everyone knows the rules of the road and will act responsibly, because, if we didn't have trust, none of us would drive, would we? Everyone who goes over a bridge, without realizing it for the most part, must have a profound belief in the skill of the individuals who have built the bridge and of those who are maintaining it. The same holds true with any human-made thing. We are interdependent, on each other, on the things we have created, and on nature.

It is doubtful that anyone wonders or cares about the race, culture, or religion of those who drive around us or those building our bridges, houses, and elevators. In the United States where many cultures and religious traditions live in communities together, work together, and pray together, race, culture, and religion really make no difference. Without ever knowing other individuals we *do* trust them and have faith in them. It is long past time for a converging of different cultures. It is time for a collapsing of values into one overarching belief. Borrowing Caroline Myss's words, it is time for us to "think in a vocabulary of oneness."[13]

Viewing energy as *one* phenomenon explains why all things are relative, why everything complements everything else, why uncertainty and unpredictability underlie our reality. It's because the one is always changing, always becoming, never standing still. How could any one thing be absolute, certain, or predictable when it never stands still? Perhaps we've set our goals too high, for it would seem that the *one* is ultimately unknowable.

For many, no doubt the question will still remain, Is there purpose in this becoming? I do believe that there have been enough intuitive thoughts expressed by philosophers and scientists in the history of humankind to warrant saying there likely is a reason for our being here. Davies says it a little more forcefully than I: "We are truly meant to be here."[14] There is an implicit intention in the way that energy, as one phenomenon, continually transforms either by nature or by our own hand in collaboration with nature. Intentionality *is* purposefulness. Look around you. Every tangible and intangible thing is a manifestation of energy, one thing that is continually changing and yet never lost. The teleological signs are all around us.

Energy is by *itself* actualizing potentials. Yes, we have found many innovative ways to use various forms of energy, in addition to those naturally supplied. As a complex life-energy form, we are capable of intentionally grasping various forms of energy that are within our reach and of transforming them in order to further actualize our potentials. What we have accomplished to date is truly incredible. And, above and beyond the uses of various environmental energy forms, we each have within (and without) ourselves the potential to transform our individual energy as the need or the will arises.

Traditional mechanical physics has encouraged us to think of energy as the ability to do work or as a source such as fuel. We certainly don't think of the inherent ability of this phenomenon to perform work or be active. Yet everything in our existence has this ability, either by nature or through humankind. Or should I say by the hand of God? Herein lies the greatest enigma.

Writing of human nature, George Leonard notes that there is "a silent pulse of perfect rhythm" within each of us, that while unique, still connects us with the rest of the universe. The evident patterns and rhythms throughout the cosmos indicate a unity, but perhaps more importantly, they imply purpose. Leonard poetically states:

> Ceaselessly pulsing, the universe expands. Stars burn out; galaxies spin more slowly.... Yet paradoxically, the universe as a whole is constantly generating new information, new order, new complexity.... A million fruit flies die to bring forth a single viable new strain.[15]

And so many of us may ask ourselves why there is such orderliness? Why, for instance, do hydrogen and oxygen so often combine to form other things? Yes, there is a chemical affinity, but the question remains, why? Why are there laws of nature, which seem to provide such an orderly universe, along with very orderly animate and inanimate forms?

We may not be able to accurately predict the future development and evolution of the universe and humankind, but I think we can be reasonably certain that there is a *reason for it all.* Even the novel, unpredictable things emerging in nature reflect purposiveness. It's simply all *much too perfect.*

Beyond a Piecemeal View

Describing the reductionistic way of science, Davies gives MacKay's illustration of an electric sign:

> ...the operation of one of those familiar advertising displays consisting of hundreds of electric lamps that flash on and off ...to spell a message. An electrical engineer could give a complete and accurate description of this system in terms of electric circuit theory, explaining exactly why and how each light is flashing. Yet the claim that the advertising display is therefore nothing but electrical pulses in a complex circuit is absurd. True the electrical description is neither wrong nor incomplete at its own particular level of description, but it makes no mention of the message.[16]

The same holds true with the phenomenon of energy. In each scientific discipline we have dissected and examined only one piece of the puzzle. Each discipline can give a complete, distinct description of energy as a particular process, form, or thing. Yet, the claim that energy is specifically this, or that, is absurd.

Physicists continue in their search for a theory of everything that will explain the entire universe. According to Hawking, "Now if you believe that the universe is not arbitrary, but is governed by definite laws, you ultimately have to combine the partial theories into a complete unified theory that will describe everything in the universe."[17] John Boslough also comments on the search for a unifying theory: "Yet a unification theory remains an almost religious vision to some scientists, a Zen-like view of reality in which all the forces and all the matter in nature come from a single source."[18] That is precisely what the idea of energy as a phenomenon seems to imply, that all the forces and all the matter in nature do indeed come from a single source.

In all its multiplicity of manifestations, energy has for some reason, unknown to us, chosen to create an orderly pattern that has formed this vast universe and all of its readily discernible and not so discernible things. That includes us and of course our innate abilities to think, to reason, to feel, to love and be loved, and to recognize the awesome beauty of it all.

In health care many individuals still hold a piecemeal view of energy (the mechanistic-reductionistic views). Conceivably, this may be necessary in order to continue to progress in certain areas of health care, particularly in areas where emphasis is placed on the physiological aspects of life forms. But there are also those who view health and well-being as a dynamic, communal process in which we all participate with the rest of the cosmos in actualizing unique potentials. It is likely that as we shed our piecemeal view of the universe, human beings will become increasingly aware of our ability to actualize innate potentials and self-healing will become ever more prevalent.

As we move into the next millennium, we will need to cautiously examine certain issues, such as DNA testing and genetic treatments — lest we make a mistake similar to treating certain infections in cancer patients when perhaps the infection is a restorative transformation for healing. How will we know that DNA testing and treatments will be the right choice we make? Rogers speculated that perhaps increasing diversity of human environmental fields is somehow preparing us, on a pandimensional level, for life in outer space. Maybe genetic mutations are somehow meaningful energy transformations. And if we do move ahead with DNA tests and treatments, will we be responsible enough not to use our knowledge of DNA testing to hurt others who are essentially a part of us?

Will cloning and computer technology lead to usurping the jobs of those who need them? Can we do what's right, despite our advancing technology? Can we shed our piecemeal view? In the TV show *Stargate,* earthlings are often referred to as "children" by the far more advanced alien life forms. The general consensus of the advanced species on *Stargate* seems to be that we earthlings need to be *protected from ourselves.* Perhaps they are right; maybe our technology should be curtailed until such time when we can act responsibly in the best interests of the *one.* The earth, the people, the other animate as well as inanimate forms, the atmosphere, the entire universe — *everything each of us does, both knowingly and unknowingly, does and will continue to contribute to the well-being or destruction of the One.*

It is widely appealing to think that all is one. Equally appealing, as the multitude of religions does indeed attest, is the idea that *we are here on purpose.* This statement, however, needs to go a step further to be congruent with the ideas that have arisen in this inquiry. One needs to say that everything exists on purpose, and it really takes little imagination to do so. There *is* a scientific basis for us to believe that there is purpose in the universe. Look at what the *one* thing has become thus far! Look at the organization and pattern all around us and in us. Watch the sunset, the sky going from blue to pink, orange, and gold. Look at the caterpillar that transforms into a butterfly. Look at us, with our ability to love, to feel, to dream. How else could the one become what it has, if not for some purpose?

Carl Sagan's closing words in his wonderful novel *Contact* poetically echo my thoughts:

> Hiding in the alternating patterns of digits, deep inside the transcendental number, was a perfect circle, its form traced out by unities in a field of noughts. The universe was made on purpose, the circle said. In whatever galaxy you happen to find yourself, you take the circumference of a circle, divide it by its diameter, measure closely enough, and uncover a miracle — another circle, drawn kilometers downstream of the decimal point. There would be richer messages farther in.... As long as you live in this universe,... sooner or later you'll find it.

> It's already here. It's inside everything.... In the fabric of space and in the nature of matter, as in a great work of art, there is, written small, the artist's signature. Standing over humans, gods, and demons... there is an intelligence that antedates the universe.[19]

Closing Thoughts

As I worked on this topic I became increasingly aware of a feeling of awe, finally leading me to the realization that no matter how complete my thoughts appear to be regarding this enigma, there will never be complete closure. To achieve it would take more than my lifetime. For as fast as we can read and write about one area relative to this enigma, another new piece of the puzzle becomes evident. Sometimes a piece of the puzzle usurps the place of the previous one or two pieces and sometimes it doesn't.

While researching the concept of energy in philosophy, physics, and the human health sciences, I kept seeing gaps, things that seemed to have been overlooked or left out. It finally dawned on me that there were no real gaps or omissions; certain things were simply not pertinent at the time they were examined or developed. For despite the subtle patterns and rhythmicities that have been discovered in the history of various sciences, we continue to have a piecemeal view of the universe.

It has been said that the ancient Chinese focused on clusterings or coincidences occurring within a time frame rather than on what we deem as causal relationships.[20] During my research a distinct feeling kept recurring: that many, many others had been saying the same thing using different words and concepts. If we look back over human history, not only does the simultaneous discovery of many important scientific theories become apparent, but the repetitive, circular way we come to rest on old ideas as if they were new is also quite evident. So, if there are those who think I have omitted pertinent ideas, then I will look upon this work as a success, for it has met the goal that I set forth in the beginning. Inquiring minds will begin to see a coherency where before there was none. You, the reader, may fill in some of the gaps of this enigma.

I can appeal, as an afterthought, to the abundance of different philosophies concerning the universe and life. As Bergson artfully notes: "There is not, there cannot be *a* philosophy as there is *a* science; on the contrary there will always be as many different philosophies as there are original thinkers."[21] But in this venture into metaphysical ideas that are linked with one of the most important scientific concepts throughout history, energy, I think the relation between science and metaphysics becomes clearer. Bergson wrote of this strange relationship where although they "differ in object and method" science and metaphysics "will commune in experience":

> The mathematical form which physics has taken is . . . what best corresponds to reality and what is most satisfying to our understanding. Much less convenient will be the position of the true metaphysics. It also will begin by eliminating ready-made concepts; it will also rely upon experience. But that inner experience of which we speak will nowhere find a strictly appropriate language. It will of course be compelled to return to the concept, with at most the addition of the image; but then it will have to enlarge the concept, make it more flexible, and indicate by the colored shading around the edge, that it does not contain the whole of experience.[22]

We are to a large extent limited by our language, and conceptual definitions will not ever capture the full essence of this enigma. As Bohr had said, "Language has this strange, fluid character. We never know what a word means exactly, and the meaning of our words depends on the way we join them together into a sentence."[23] At best, all we can do is continue to refine and clarify our language to more accurately reflect our growing knowledge of the cosmos. But our hearts, our minds, and, most significant, our experiences will always lend truth to our individual convictions, whatever they may be.

If after all this one asks, What is the enigma? I would answer: it's the fact that every spring the flowers bloom and the trees turn green again; that I wake up to birds singing in the morning; that the sky appears so blue; that the stars are so bright and that they are in a real sense my ancestors; that there is a discernible pattern in every snowflake; that plants (and humans) naturally turn toward the sunlight; that a song can elicit feelings of immense joy or sorrow;

that I can choose my future and actualize my unique potentials — as can every human being.

It's the things we've created thus far, like the automobile, the fax machine, the Pentium computer, the telegram, and the planes that fly. Penicillin, aspirin, and interferon. . . . The enigma is everything, and everything is energy. It's all one gigantic whole. And to know that is to know that it all couldn't possibly *just happen to be here.* The very fact that it is and that we humans can make any sense of it — the physical laws, the philosophical arguments, and all the other knowledge we've managed to accumulate — points to some intention, some purpose, some *reason for it all.*

Notes

Introduction

1. Phillips, "Search in Research."
2. Moran, *A Grammar of Responsibility,* 1.
3. Bergson, *The Creative Mind.*
4. Ibid., 110–11.
5. Ibid., 111.
6. Horgan, "Science Set Free from Truth," A17.

Chapter 1: The Enigma

1. Emerson, *The Works of Ralph Waldo Emerson,* 541.
2. Johnson, "Researchers on Complexity Ponder What Exactly They Study," C1, C7.
3. Van Ness, *Understanding Thermodynamics,* 2.
4. Rogers, *An Introduction to the Theoretical Basis of Nursing.*
5. Csikszentmihalyi, *Flow: The Psychology of Optimal Experience; Creativity: Flow and the Psychology of Discovery and Invention;* Maslow, *Toward a Psychology of Being.*
6. See Barrett, "A Nursing Theory of Power for Nursing Practice."
7. Selye, *The Stress of Life,* 52.
8. Feldman, "Energy," 123.
9. See Einstein, *Ideas and Opinions.*
10. Ibid., 277.
11. See Capra, *The Tao of Physics.*
12. Chan, *A Source Book in Chinese Philosophy,* 784, notes: "Every student of Chinese thought knows that *ch'i* as opposed to *li* (principle) means both energy and matter, a distinction not made in Chinese philosophy. Both 'matter' and 'ether' are inadequate. Dubs' 'matter-energy' is essentially sound but awkward and lacks an adjective form. Unless one prefers transliteration, 'material force' seems to be the best. In many cases, especially before the Neo-Confucian doctrine of *li* developed, *ch'i* denotes the psychophysiological power associated with blood and breath. As such it is translated as 'vital force' or 'vital power.' " See also Ivanhoe, "Ch'i" and "li" in *The Cambridge Dictionary of Philosophy.*
13. Lao Tzu, *Tao Te Ching.* See also Lusthaus, "Laozi (Lao Tzu)."

14. Sellmann, "Zhang Zai (Chang Tsai)," 108.

15. Radhakrishnan and Moore, *A Sourcebook in Indian Philosophy,* xxv.

16. Zysk, "Traditional Ayurveda," In *Fundamentals of Complementary and Alternative Medicine.*

17. See Barnes, *The Pre-Socratic Philosophers* and Guthrie, *The Greek Philosophers.*

18. Lindsay, *Energy,* edited a text in which he compiled a number of seminal works believed to be related to the development of the concept of energy. Lindsay claims that Aristotle's *energeia* bears no resemblance to the modern conception of energy vis-à-vis constancy in the midst of change. But he notes also that in Aristotle's *Physics,* Book VIII, there were traces of the energy idea. See also Lindsay, *Applications of Energy,* 298–307.

19. Lindsay, *Energy,* 44. See also Parmenides, *Parmenides: A Text with Translation, Commentary and Critical Essays.*

20. Lucretius, *Of the Nature of Things,* 8.

21. Ibid., 12.

22. Turnbull, *The Forest People.*

23. See Jackomos and Fowell, *Aboriginal Living Histories.*

Chapter 2: A Very Brief History of Energy

1. Selye, *The Stress of Life,* 42.

2. Sagan, *Contact,* 178.

3. West, *The World Is Made of Glass,* 47.

4. See Jammer, "Energy," for a history of the concept.

5. See Ackrill, "Aristotle's Distinction between *Energeia* and *Kinesis,*" for a discussion of *energeia* in relation to *kinesis.*

6. Aristotle, *The Metaphysics,* Book IX, 329, Bekker code: *Metaphysics* IX 8, 1049b4.

7. Aristotle, *The Complete Works of Aristotle,* 1603, Bekker code: *Metaphysics* IV 4, 1015a17–19.

8. The tendency has often been to replace the word "stuff" with that of "substance" when in fact Aristotle draws a distinction between the two terms.

9. See Yoon, "Clues to Redwood's Mighty Growth Emerge."

10. See Gill, *Aristotle on Substance,* for an in-depth examination of Aristotle's concept of "substance."

11. Lear, *Aristotle,* 19–20.

12. Aristotle, *The Complete Works of Aristotle,* 1658, Bekker code: *Metaphysics* IX 6, 1050a 6–10.

13. See Jammer, "Energy."

14. Maimonides, *The Essential Maimonides.*

15. Aquinas, *The Summa Theologica,* 25.

16. Descartes, *Principles of Philosophy,* in *The Philosophical Works of Descartes,* 265.

17. Ibid., 267. See also Hick, *The Existence of God.*

18. Newton, *The Principia.*

19. Leibniz, *The Monadology,* 222.

20. Ibid., 228.

21. Latta (*Leibniz,* 229) notes, "The entelechy of Leibniz . . . is to be understood as an individual substance or force, containing within itself the principle of its own changes. It is called entelechy, not because it is a state of perfect realization, but because it contains in germ an infinity of perfections, which it tends to develop. It is . . . not so much the final developed condition of a thing, opposed to its potentiality, but rather it implies the tendency or virtuality . . . as something intermediate between the bare potency and the fully developed activity." He also notes: "*Entelecheia* in Aristotle is the *state* of perfection or realization in which *energeia,* as a *process* ends."

22. See Bernoulli, in "Jean Bernoulli's Principle of Virtual Velocities"; Jammer, "Energy."

23. See Kuhn, "Energy Conservation as Simultaneous Discovery."

24. Mayer, "On the Forces of Inorganic Nature," 68–69.

25. Joule, "On Matter, Living Force, and Heat," 350.

26. Colding, "On the History of the Principle of the Conservation of Energy," 160.

27. Colding, "Scientific Reflections on the Relationship between Intellectual Life's Activity and the General Forces of Nature," 120–21.

28. Reference is made to Cabeo (1586–1650), who apparently conceived the notion of lines of force as early as 1629. See May, "Lines of Force."

29. See Harman, *Energy, Force, and Matter.*

30. Faraday, *The Forces of Matter,* 87.

31. See Einstein and Infeld, *The Evolution of Physics.*

32. Maxwell, *A Dynamical Theory of the Electromagnetic Field,* 70.

33. Poynting, "On the Transfer of Energy in the Electromagnetic Field," 516.

34. Lewen, "Punctuated Equilibrium Is Now Old Hat," 672.

35. Feynman, *Six Easy Pieces,* 71.

36. G. DiPego (writer), B. Boyle and M. Taylor (producers), J. Turteltaub (director), 1966. Available from Touchstone Home Videos, distributed by Buena Vista Home Video, Burbank, Calif.

37. Johnson, "Once upon a Time, There Was a Big Bang Theory"; "Elusive Particles Continue to Puzzle Theorists of the Sun." In addition, see Browne, "Neutrinos Have Mass, but Not Enough to Hold the Universe Together"; "Mass Found in Elusive Particle; Universe May Never Be the Same"; "Room in the Universe for Ancient Beliefs and Modern Physics"; Gibbs, "A Massive Discovery"; Normile, "Weighing in on Neutrino Mass"; Singh, "The Proof Is in the Neutrino."

38. Normile, "Weighing in on Neutrino Mass," 1690.

39. Haisch, Rueda, and Puthoff, "Beyond $E=mc^2$," 26.

40. Ibid.

41. Ibid., 30

42. Hesse, "Ether," discusses the origin of the concept of aether. See also Reichenbach (*From Copernicus to Einstein,* who discusses the concept of ether from the time of Newton.

43. See Newton, *Newton's Optical Writings,* particularly queries 18–24 in *The Opticks,* Book III.

44. Lodge, *Energy,* 97.
45. Ibid.
46. Einstein and Infeld, *The Evolution of Physics,* 176.
47. Planck, *The Universe in the Light of Modern Physics,* 45–46.
48. Heisenberg, *Physics and Philosophy,* 41.
49. Ibid., 42–43.
50. Bohr, *Essays 1958–1962 on Atomic Physics and Human Knowledge.*
51. Johnson, "The Unspeakable Things That Particles Do," E5.
52. Browne, "Physicists Put Atom in 2 Places at Once," C1.
53. Monroe, Meekhof, King, and Wineland, "A 'Schrodinger Cat' Superposition State of an Atom."
54. Briggs and Peat, *Looking Glass Universe,* 49.
55. Davies, *The Cosmic Blueprint,* 7.
56. V. Salva (writer and director), R. Birnbaum and D. Grodnik (producers), 1996. Available from Hollywood Pictures, Burbank, Calif.

Chapter 3: The Becoming of It All

1. Davies, *The Cosmic Blueprint,* 3.
2. Russell, *The Awakening Earth,* 40. There are a number of excellent books that discuss the big bang and other theories relative to the creation of the universe. For example, see Gribbin, *In Search of the Big Bang,* and Weinberg, *The First Three Minutes.* Also see Davies; a partial listing of his books is provided in the bibliography.
3. Hawking, *A Brief History of Time,* 70.
4. Linde, "The Self-Reproducing Universe," 102.
5. Ibid., 104.
6. As I am wrapping things up for this book, scientists in Italy think that they have detected a "dark particle" with sixty times more mass than a proton — WIMPS — and that these particles may be able to account for a lot of the missing stuff. Of course, it isn't definitive yet. See Glanz, "Has a Dark Particle Come to Light?"
7. See Wilford, "In the Light of Dying Stars, Astronomers See Intimations of Cosmic Immortality."
8. See Wilford, "Cosmologists Ponder 'Missing Energy' of the Universe."
9. Overbye, "A Famous Einstein 'Fudge' Returns to Haunt Cosmology," 1.
10. For more on cosmic theories and the missing stuff see Browne, "Neutrinos Have Mass, but Not Enough to Hold the Universe Together"; "Room in the Universe for Ancient Beliefs and Modern Physics"; Bucher and Spergel, "Inflation in a Low-Density Universe"; Glanz, "Exploding Stars Point to a Universal Repulsive Force" and "Astronomers See a Cosmic Antigravity Force at Work"; "Has a Dark Particle Come to Light?"; Hogan, Kirshner, and Suntzeff, "Surveying Space-Time with Supernovae"; Johnson, "Once upon a Time, There Was a Big Bang Theory"; Krauss, "Cosmological Antigravity"; Overbye, "A Famous Einstein 'Fudge' Returns to Haunt Cosmology"; Watson, "Inflation Confronts an Open Universe"; Wilford, "Scientists Report Discovering 'Missing Mass'

Component"; "In the Light of Dying Stars, Astronomers See Intimations of Cosmic Immortality"; "Cosmologists Ponder 'Missing Energy' of the Universe." I'm sure there are more out there. See www.nytimes.com (the science archives).

11. Thuan, *The Birth of the Universe,* 72.

12. Davies, *The Cosmic Blueprint,* 127.

13. Leslie, *Physical Cosmology and Philosophy,* 3.

14. See Barrow and Tipler, *The Anthropic Cosmological Principle.*

15. Microwave background radiation is radiation from the very early stages of the universe which can be picked up as sound by a very sensitive microwave detector. See Hawking, *A Brief History of Time.*

16. See Jastrow, *Red Giants and White Dwarfs.* Also see Thuan, *The Birth of the Universe,* for a colorful depiction of cosmic evolution.

17. Hawking, *A Brief History of Time,* 112.

18. Pellegrino and Stoff, *Darwin's Universe,* x.

19. See Wilson, *The Diversity of Life.*

20. See Yoon, "Lake Victoria's Lightning-Fast Origin of Species." Lake Victoria in Africa, which is only twelve thousand years old, has over six hundred different, although obviously related, species of cichlid fish. For some reason these numerous species of fish have evolved in the short space of time that Lake Victoria has existed.

21. Darwin, *The Origin of Species,* 404.

22. Thomas Huxley reprinted in Himmelfarb, *Darwin and the Darwinian Revolution,* 343–44.

23. Schlegel, "Time and Thermodynamics," 505.

24. Ibid., 506.

25. For an enlightening book on chaos theory see Gleick, *Chaos.*

26. Lorenz, *The Essence of Chaos,* 8.

27. Ibid., 181–82.

28. See ibid. for further elaboration.

29. Kauffman, *At Home in the Universe,* 283.

30. See Prigogine and Stengers, *Order Out of Chaos.* Also see Prigogine, *From Being to Becoming,* and *The End of Certainty.*

31. Dissipative structures are also called autopoietic, from the Greek for "self-production." See Jantsch, *The Self-Organizing Universe.*

32. Ibid., 40.

33. Bohm, *Wholeness and the Implicate Order,* 159.

34. Talbot, *The Holographic Universe,* 52.

35. See Bohm and Peat, *Science, Order, and Creativity.* They also discuss two other orders: the generative and superimplicate orders. The generative order is involved with creative processes while the superimplicate order, occasionally referred to as the "superfield," is said to guide and organize the original field (the first implicate order).

36. Bohm, *Wholeness and the Implicate Order,* 177.

37. Ibid., 151.

38. Mitchell, *The Way of the Explorer,* 165.

39. Ibid., 134. See Haisch et al., "Beyond $E=mc^2$"; they also discuss the zero-point energy field. See also Bohm, *Wholeness and the Implicate Order.*

40. Woodhouse, *Paradigm Wars,* 161.
41. Ibid., 187.
42. Ibid., 183.
43. Ibid., 164.
44. See Burr, *The Fields of Life;* Russell, "The Fields of Life."
45. Sheldrake, "Cause and Effect in Science: A Fresh Look," 79. See also Sheldrake, *A New Science of Life.*
46. Sheldrake, "Cause and Effect in Science," 81. Morphic resonance is also referred to by Sheldrake as formative causation.
47. Jung, *The Archetypes and the Collective Unconscious,* 4.
48. Lovelock, *Gaia.*
49. Polkinghorne, *Beyond Science,* 122.
50. See Browne, "Granular Objects, if Jiggled, Mimic Atoms and Molecules."
51. Jantsch, *The Self-Organizing Universe,* 35.
52. Prigogine and Stengers, *Order Out of Chaos,* 9.
53. Einasto et al., "A 120-Mpc Periodicity in the Three-Dimensional Distribution of Galaxy Superclusters."
54. Johnson, "The Real Star Wars: Between Order and Chaos," E 4.
55. Sheldrake, "Cause and Effect in Science," 81.
56. See Gadamer, *The Enigma of Health.*

Chapter 4: Human Beings: More Than Just Matter

1. Keen, *I Believe in God and in Evolution,* 100.
2. Taylor notes: "*Psyche* or 'soul' means in Greek more, and less, than 'consciousness' does to us. Consciousness is a relatively late and highly developed manifestation of the principle which the Greeks call 'soul.'...Thus consciousness is more a secondary feature of the 'soul' in Greek philosophy than in modern thought" (*Aristotle,* 75). In Plato's works, the soul has three parts; mind, spirit, and desire.
3. Plato, *Timaeus,* in *Dialogues of Plato,* 374.
4. Aristotle, *The Complete Works of Aristotle.*
5. Aristotle as quoted by Barnes, *Aristotle,* 68.
6. Descartes, *Principles of Philosophy,* in *The Philosophical Works of Descartes,* 221.
7. Descartes, *Discourse on Method and Meditations,* 132.
8. Ibid., 134.
9. Leibniz, *The Monadology,* 257. His view, as well as Teilhard's, falls under the auspices of panpsychism, which holds that all matter has a kind of consciousness. Whitehead, *Nature and Life,* also shared this view.
10. Leibniz, *The Monadology,* 256.
11. Ibid., 256–57.
12. Bergson, *Creative Evolution,* 27.
13. Ibid., 119.
14. Ibid., 120.
15. Teilhard de Chardin, *The Phenomenon of Man,* 63–64.

16. For an discussion of Teilhard's hyperphysics, see O'Manique, *Hyperphysics.*
17. Teilhard, *Human Energy,* 45.
18. Darling, *Soul Search,* xix.
19. Flanagan, *The Science of the Mind,* 364.
20. Ibid., 20. Flanagan also has an interesting argument against the Cartesian nonphysical mind on the basis of the law of energy conservation. He gives an example of a person deciding to go to a concert. If the person goes to the concert, he or she will have used energy, physical energy, to do so. But in order for the energy to increase in a system, which it must if one is expending it to go to the concert, the energy must have "been transferred from some other physical system" (21). But since the mind, according to Descartes, is nonphysical, the mind "does not have any energy to transfer" (21). So with the Cartesian idea of a nonphysical mind, the question arises: how did the person get to the concert if the mind did not (because it could not) transfer the energy into the action?
21. Dennett, *Consciousness Explained.* See also Wright, "Can Machines Think?"
22. Chalmers, *The Conscious Mind,* 125.
23. Ibid., 213.
24. See Wright, "Can Machines Think?"
25. Chalmers, *The Conscious Mind,* 242–43.
26. Pribram, "Some Dimensions of Remembering."
27. Penrose, *The Large, the Small and the Human Mind,* 101.
28. Theories of the mind being in the matter all fall in the realm of epiphenomenalism. They each seek to define some aspect of sentience in terms of a causal connection between physical and mental events.
29. Dossey, "Where in the World Is the Mind?" 40.
30. Ibid.
31. Sagan, *Broca's Brain.*
32. Ring, *Life at Death,* 237.
33. Ring speculated that the "being of light" reported in most NDEs could be the person's higher self. He goes on to note that one's higher self can be construed as an "aspect of God" (241). See also Ring, "Near-Death Experiences: Implications for Human Evolution and Planetary Transformation."
34. Darling, *Soul Search,* 155.
35. Ibid., 156.
36. Phillips, "Changing Human Potentials and Future Visions of Nursing," 20.
37. Moody, *Life after Life,* 128.
38. Dossey, "Where in the World Is the Mind?" 43.
39. Maslow, *Toward a Psychology of Being; The Farther Reaches of Human Nature.*
40. Margenau, "Science, Creativity, and Psi," 83.
41. Reinsel, "Parapsychology: An Empirical Science," 165. The Greek letter *psi* is often used in the current literature to represent the subject matter of parapsychology, ESP, and psychokinesis.
42. Rhine, "Extrasensory Perception," 168.

43. See Wolman, *Handbook of Parapsychology.*

44. Taylor, *Superminds,* 119.

45. Jung is said to have derived his ideas from the Austrian biologist Paul Kammerer. See Koestler, *The Roots of Coincidence.*

46. Jung, *The Portable Jung,* 518.

47. Ibid., 505.

48. Many people die on or about the time of their birthdays.

49. Jung, *Synchronicity,* 19.

50. Peat, *Synchronicity,* 17.

51. See also Koestler, *The Roots of Coincidence.*

52. White and Krippner, *Future Science,* 550.

53. Ibid., 53–54.

54. See Dossey, "But Is It Energy?"; "Healing, Energy, and Consciousness"; "Dialogue: But Is It Energy?"; "Energy Talk"; "The Forces of Healing."See also Leskowitz, "Dialogue"; Tiller, "Dialogue: But Is It Energy?"

55. Dossey, "But Is It Energy?," 75.

56. Ibid., 75. See also Herbert, *Quantum Reality.*

57. Dossey, "Healing, Energy, and Consciousness," 14.

58. For more information on the science of unitary human beings see Rogers, *An Introduction to the Theoretical Basis of Nursing;* "Nursing: A Science of Unitary Man"; "A Science of Unitary Human Beings"; "Nursing Science and the Space Age." Also see Barrett, *Visions of Rogers' Science-Based Nursing;* Madrid, *Patterns of Rogerian Knowing;* Madrid and Barrett, *Rogers' Scientific Art of Nursing Practice;* Malinski, *Explorations on Martha Rogers' Science of Unitary Human Beings.*

59. See Parse, *Major Paradigms, Theories, and Critiques;* Fawcett, "From a Plethora of Paradigms to Parsimony in Worldviews"; "On the Requirements for a Metaparadigm," and Newman, "Prevailing Paradigms in Nursing," for discussions on different paradigmatic views in nursing. See also Butcher, "In Search of Sympathetic Vibrations," who compares and contrasts Rogerian views with the theories of Bohm, Prigogine, and Sheldrake.

60. Phillips, "Changing Human Potentials and Future Visions of Nursing," 17.

61. LeShan, *The Medium, the Mystic, and the Physicist,* 48.

62. See Andersen and Smereck, "Personalized Nursing LIGHT Model"; Sarter, *The Stream of Becoming;* "Some Critical Philosophical Issues in the Science of Unitary Human Beings"; Todaro-Franceschi, "Must Acausality Imply Unpredictability?"

63. Schrodinger, *What Is Life? With Mind and Matter,* 127.

64. As cited by Darling, *Soul Search,* 58.

Chapter 5: The Nature of Health and Healing

1. Nightingale, *Notes on Nursing,* 5.

2. Gadamer, *The Enigma of Health,* 107.

3. For a general treatment of the history of ancient medicine, see Magner, *A*

History of Medicine; Seelig, *Medicine: An Historical Outline;* Sigerist, *A History of Medicine.*

4. See Dossey, *Healing Words.*

5. Hippocrates, "Epidemics," 255. See also Hippocrates, "Ancient Medicine" and "Airs, Waters, Places," in *The Medical Works of Hippocrates.*

6. Inglis, *A History of Medicine,* 3–4.

7. Nightingale, *Notes on Nursing,* 5.

8. Ibid., 49.

9. See also Nightingale, *Suggestions for Thought,* and Macrae, "Nightingale's Spiritual Philosophy."

10. Nightingale, *Notes on Nursing,* 74–75.

11. For example, see Selye, *The Stress of Life,* and Levine, "The Four Conservation Principles of Nursing"; "The Conservation Principles: A Retrospective."

12. Weil, *Spontaneous Healing,* 38–39.

13. Caplan, "The Clash over Quackery," 23.

14. Inglis, *A History of Medicine,* 185.

15. Slater, "Toward an Understanding of Energetic Healing," 227.

16. Aristotle would probably not consider a wave function an actuality at all.

17. Kaptchuk, "Historical Context of the Concept of Vitalism," 35.

18. See Burr, *The Fields of Life,* Russell, "The Fields of Life"; Kunz and Peper, "Fields and Their Clinical Implications."

19. Seaward, "Alternative Medicine Complements Standard," 54.

20. Slater, "Toward an Understanding of Energetic Healing," part 2: "Energetic Processes," 235; see also Herbert, *Quantum Reality.*

21. See Krieger, *Accepting Your Power to Heal;* "Higher-Order Emergence of the Self during Therapeutic Touch." Dolores Krieger and Dora Kunz developed the therapeutic touch healing modality. See also Macrae, *Therapeutic Touch;* "Therapeutic Touch as Meditation."

22. See Krieger, *Accepting Your Power to Heal;* Kunz, *Spiritual Healing;* Macrae, *Therapeutic Touch;* Meehan, "Therapeutic Touch and Postoperative Pain"; Quinn, "Therapeutic Touch as Energy Exchange"; "Therapeutic Touch as Energy Exchange: Replication and Extension."

23. Talbot, *The Holographic Universe,* 188.

24. Slater, "Toward an Understanding of Energetic Healing," part 1: "Energetic Structures."

25. O'Regan, "Healing, Remission and Miracle Cures," 46.

26. Ibid., 48.

27. Dossey, *Healing Words,* 64.

28. See O'Regan, "Healing, Remission and Miracle Cures."

29. Rogers developed three key principles of homeodynamics to explain the nature of human and environmental change: the principles of helicy, resonancy, and integrality. The principle of helicy posits that change is continuously innovative and unpredictable and that there is an increasing diversity of the human and environmental fields. The principle of resonancy asserts that wave patterns in human and environmental energy fields are unceasingly changing from lower to higher frequencies. The principle of integrality denotes the perpetual whole-

ness of the human environmental energy fields, which are infinitely in a mutual process of innovative becoming.

30. See Madrid, "Participating in the Process of Dying"; McEvoy, "The Relationships among the Experiences of Dying, the Experience of Paranormal Events, and Creativity in Adults"; Phillips, "Changing Human Potentials and Future Visions of Nursing."

31. Between B cells, T cells, macrophages, and RNA, etc.

32. Barrett, "A Nursing Theory of Power for Nursing Practice," 207. See also Barrett, Caroselli, Smith and Smith, "Power as Knowing Participation in Change."

33. Field, "The Scientific Art of Medical Practice," 269.

34. See Dossey, *Healing Words.*

35. There are currently six centers (called Star Centers) located throughout the United States that teach actualism techniques. Dr. Bruce Jaffe and Bernice B. Cousins are co-directors of the New York Center; both are ordained ministers. Web site: http://www.actualism.org. Dr. Jaffe can be contacted directly at brjaffe@actualism.org.

36. James, *William James,* 1227.

37. Macrae, *Therapeutic Touch,* 18.

38. Phillips, "Changing Human Potentials and Future Visions of Nursing."

39. Gold, "Practicing Medicine in the Nineties with an Emphasis on the Unitary Perspective of Patient Care," 266.

40. Dalai Lama and Cutler, *The Art of Happiness,* 201.

Chapter 6: The Spiritual Essence of the Enigma

1. Teilhard de Chardin, *The Divine Milieu,* 114.

2. Capra and Steindl-Rast, *Belonging to the Universe.*

3. See Capra, *The Tao of Physics;* Smart, "Hinduism"; also McGreal, *Great Thinkers of the Eastern World,* for an in-depth description of Eastern philosophy.

4. Weber, *Dialogues with Scientists and Sages,* 13.

5. Dalai Lama in ibid., 242.

6. Permission to quote from the Science and Spiritual Quest conference abstract (June 1998) was graciously given by Dr. Rudd via e-mail.

7. Whitehead, "The New Reformation," 217.

8. Bohm, *On Creativity,* 28.

9. Hildegard of Bingen, as quoted in Fox, *Illuminations of Hildegard of Bingen.*

10. See Parmenides, *Parmenides: A Text with Translation, Commentary and Critical Essays.* At least the argument has been that if one thinks of some "thing," even to think of "nothing" means that "nothing" exists. For how could any human mind think of nothing?

11. Singh, "The Proof Is in the Neutrino."

12. See chapters 2 and 3.

13. Aquinas's disagreement with Aristotle's conception of change is expressed in *The Summa Theologica,* Part one, q. 46, art. 1–2.

14. When I refer to "laws of nature" I am not referring to man-made laws; rather, I am referring to what natural philosophers have called "the laws of nature."

15. Prigogine, *The End of Certainty,* 6.

16. Phillips, "The Open-Ended Nature of the Science of Unitary Human Beings," 21.

17. Davies, *About Time,* and Fraser, *The Voices of Time,* touch upon some of the major issues regarding time, including the idea believed by some people that there are multiple realities.

18. Eckhart as quoted by LeShan, *The Medium, the Mystic, and the Physicist,* 49.

19. In addition to Davies, *About Time,* I refer the reader once again to Hawking, *A Brief History of Time.*

20. Davies, *About Time,* 141.

21. Cooper, *God Is a Verb,* 38.

22. See Teilhard de Chardin, *The Heart of Matter.*

23. Eckhart, *Meister Eckhart,* 259.

24. The most recent article that I'm aware of was in the *New York Times,* November 24, 1998 (see Christensen, "Teachers Fight for Darwin's Place in U.S. Classroom"). Over 150 years have passed since Darwin's theory of the origin of the species was published and teachers are still fighting for the right to teach evolutionary theory in various schools in these United States. The argument for *not* teaching evolutionary theory is based on the idea that it is only theory and not fact. But even accepting this thesis, we really should note that all so-called "fact" is based on theory, some of which has emerged solely from intuition.

25. Dawkins, *The Blind Watchmaker,* 317.

26. Ibid.

27. Ibid.

28. Paley, "The Watch Analogy," 85.

29. Polanyi, *Personal Knowledge.*

30. See also Gallagher ("'Natural Selection': A Tautology?"), who notes that Darwin's theory is meant (by the characterization of the process of variation as "random") to deny, not claim, teleological explanation.

31. Aristotle, *Physics,* 50–51, Bekker code: *Physics* II 8, 198b24–31.

32. Collingwood, *The Idea of Nature,* 95.

33. Ibid., 5.

34. See Capra and Steindl-Rast, *Belonging to the Universe.*

35. Collingwood, *The Idea of Nature,* 146.

36. Prigogine, as quoted by Dossey, *Space, Time and Medicine,* 82.

37. Capra and Steindl-Rast, *Belonging to the Universe,* 91.

38. Angelus Silesius as translated by Harvey, *The Essential Mystics,* 207–8.

39. See Moran, *A Grammar of Responsibility.*

40. Capra and Steindl-Rast, *Belonging to the Universe,* 24.

41. Increasing human diversity is brought to light particularly in the literature encompassing parapsychology and Rogerian views.

42. Bucke, *Cosmic Consciousness,* 61.

43. Watson, *Supernature,* 315.

44. Toolan, "Nature Is a Heraclitean Fire," 31. Appreciation is extended to the Seminar on Jesuit Spirituality for permission to quote Father Toolan at length.
45. Cooper, *God Is a Verb,* 98.
46. In Capra and Steindl-Rast, *Belonging to the Universe,* 88.
47. Ibid., 88.
48. Henslow, as reprinted by Himmelfarb, *Darwin and the Darwinian Revolution,* 313.
49. Toolan, "Nature Is a Heraclitean Fire," 43.
50. Underhill, "Mysticism and Theology," 526.
51. Berger, *A Rumor of Angels,* 7.
52. Johnson, *Fire in the Mind,* 328.
53. Hawking, *A Brief History of Time,* 122.
54. Ibid; also see Davies, *The Mind of God.*

Chapter 7: Full Circle: Back to the Enigma

1. Einstein, *Ideas and Opinions,* 11.
2. Teilhard, *The Heart of Matter,* 198.
3. Zukav, *The Dancing Wu Li Masters,* 134.
4. Ackrill, "Aristotle's Distinction between *Energeia* and *Kinesis,*" 121.
5. West, *The World Is Made of Glass,* 47.
6. Peat, *Synchronicity,* 53.
7. Ibid., 55.
8. Briggs and Peat, *Looking Glass Universe,* 271.
9. Zukav, *The Dancing Wu Li Masters,* 297.
10. Hume, *On Human Nature and the Understanding.*
11. Ibid., 80–81.
12. Bohm, *On Creativity,* 110.
13. Myss, *Anatomy of the Spirit.*
14. Davies, *The Mind of God,* 232.
15. Leonard, *The Silent Pulse,* xii, 178.
16. Davies, *God and the New Physics,* 62.
17. Hawking, *A Brief History of Time,* 12.
18. Boslough, *Stephen Hawking's Universe,* 5–6.
19. Sagan, *Contact,* 430–31.
20. Peat, *Synchronicity.*
21. Bergson, *The Creative Mind,* 133.
22. Ibid., 131–32.
23. Bohr, as quoted by Heisenberg, *Physics and Beyond,* 134.

Bibliography

Ackrill, J. L. "Aristotle's Distinction between *Energeia* and *Kinesis*." In *New Essays on Plato and Aristotle,* ed. R. Bambrough, 121–41. New York: Humanities Press, 1965.

Andersen, M. D., and G. A. D. Smereck. "Personalized Nursing LIGHT Model." *Nursing Science Quarterly* 2 (1989): 120–30.

Aquinas, St. T. *The Summa Theologica.* In *Introduction to St. Thomas Aquinas,* ed. A. C. Pegis. New York: Modern Library, 1948.

Aristotle. *The Complete Works of Aristotle: The Revised Oxford Translation.* Vols. 1–2. Ed. J. Barnes. Princeton, N.J.: Princeton University Press, 1984.

———. Aristotle. *The Metaphysics,* Book IX. In *A New Aristotle Reader,* ed. J. L. Ackrill, 321–35. Princeton, N.J.: Princeton University Press, 1987.

———. *Physics.* Trans. R. Waterfield. New York: Oxford University Press, 1996.

Barnes, J. *Aristotle.* New York: Oxford University Press, 1982.

———. *The Pre-Socratic Philosophers.* Rev. ed. New York: Routledge, 1993.

Barrett, E. A. M. "A Nursing Theory of Power for Nursing Practice." In *Conceptual Models for Nursing Practice,* ed. J. P. Riehl-Sisca, 207–17. 3d ed. Norwalk, Conn.: Appleton and Lange, 1989.

Barrett, E. A. M., ed. *Visions of Rogers' Science-Based Nursing.* New York: National League for Nursing, 1990.

Barrett, E. A. M., C. Caroselli, A. S. Smith, and D. W. Smith. "Power as Knowing Participation in Change: Theoretical, Practice, and Methodological Issues, Insights, and Ideas." In *Patterns of Rogerian Knowing,* ed. M. Madrid, 31–46. New York: National League for Nursing, 1997.

Barrow, J. D., and F. J. Tipler. *The Anthropic Cosmological Principle.* New York: Oxford University Press, 1986.

Berger, P. L. *A Rumor of Angels: Modern Society and the Rediscovery of the Supernatural.* New York: Doubleday, 1969.

Bergson, H. *Creative Evolution.* Trans. A. Mitchell. New York: Modern Library, 1944; originally published 1911.

———. *Selections from Bergson.* Ed. H. A. Larrabee. New York: Appleton-Century-Crofts, 1949.

———. *The Creative Mind: An Introduction to Metaphysics.* New York: Carol Publishing Group, 1992; originally published 1946.

Bernoulli, J. In "Jean Bernoulli's Principle of Virtual Velocities." In *A Source Book in Physics,* ed. F. Magie, 48–50. Cambridge, Mass.: Harvard University Press, 1963; originally published 1935.

Bohm, D. *Wholeness and the Implicate Order.* New York: Ark Paperbacks, 1980.

———. *On Creativity.* Ed. L. Nichol. New York: Routledge, 1998.

Bohm, D., and F. D. Peat. *Science, Order, and Creativity.* New York: Bantam, 1987.

Bohr, N. *Essays 1958–1962 on Atomic Physics and Human Knowledge.* Bungay, Suffolk: Richard Clay, 1963.

Boslough, J. *Stephen Hawking's Universe.* New York: Avon, 1980.

Briggs, J. P., and F. D. Peat. *Looking Glass Universe: The Emerging Science of Wholeness.* New York: Simon & Schuster, 1984.

Browne, M. W. "Neutrinos Have Mass, but Not Enough to Hold the Universe Together." *New York Times,* May 7, 1996, C7.

———. "Physicists Put Atom in 2 Places at Once." *New York Times,* May 28, 1996, C1, C5.

———. "Granular Objects, if Jiggled, Mimic Atoms and Molecules." *New York Times,* September 3, 1996, C1, C8.

———. "Mass Found in Elusive Particle; Universe May Never Be the Same." *New York Times (On the Web),* www.nytimes.com, June 5, 1998, 1–2.

———. "Room in the Universe for Ancient Beliefs and Modern Physics." *New York Times (On the Web),* www.nytimes.com, June 9, 1998, 1–3.

Bucher, M. A., and D. N. Spergel. "Inflation in a Low-Density Universe." *Scientific American* 280, no. 1 (1999): 62–69.

Bucke, R. M. *Cosmic Consciousness: A Study in the Evolution of the Human Mind.* New York: Carol Publishing Group, 1993.

Burr, H. S. *The Fields of Life.* New York: Ballantine, 1973.

Butcher, H. K. "In Search of Sympathetic Vibrations: A Response to Parse's 'Investing the Legacy: Martha E. Rogers' Voice Will Not Be Silenced." *Visions: The Journal of Rogerian Nursing Science* 5 (1997): 12–17.

Caplan, R. "The Clash over Quackery: Protecting Alternative Care." *Health/PAC Bulletin* (Winter 1987): 22–27.

Capra, F. *The Tao of Physics.* 3d ed. Boston: Shambhala, 1991.

Capra, F., and D. Steindl-Rast. with T. Matus. *Belonging to the Universe.* San Francisco: Harper & Row, 1991.

Chalmers, D. J. *The Conscious Mind: In Search of a Fundamental Theory.* New York: Oxford University Press, 1996.

Chan, Wing-Tsit. *A Source Book in Chinese Philosophy.* Princeton, N.J.: Princeton University Press, 1963.

Christensen, J. "Teachers Fight for Darwin's Place in U.S. Classroom." *New York Times (On the Web),* www.nytimes.com, November 24, 1998, 1–4.

Colding, L. "On the History of the Principle of the Conservation of Energy." In *Ludwig Colding and the Conservation of Energy Principle: Experimental and Philosophical Contributions,* trans. P. F. Dahl, 159–66. New York: Johnson Reprint Corp., 1972; originally published 1863.

———. "Scientific Reflections on the Relationship between Intellectual Life's Activity and the General Forces of Nature." In *Ludwig Colding and the Conservation of Energy Principle: Experimental and Philosophical Contributions,* trans. P. F. Dahl, 105–27. New York: Johnson Reprint Corp., 1972; originally published 1856.

Collingwood, R. G. *The Idea of Nature.* Westport, Conn.: Oxford University Press, 1944.

Cooper, D. A. *God Is a Verb: Kabbalah and the Practice of Mystical Judaism.* New York: Riverhead Books, 1997.

Csikszentmihalyi, M. *Flow: The Psychology of Optimal Experience.* New York: Harper & Row, 1990.

———. *Creativity: Flow and the Psychology of Discovery and Invention.* New York: HarperCollins, 1996.

Dalai Lama and H. C. Cutler. *The Art of Happiness.* New York: Riverhead Books, 1998.

Darling, D. *Soul Search: A Scientist Explores the Afterlife.* New York: Villard Books, 1995.

Darwin, C. *The Origin of Species.* New York: Penguin, 1958; originally published 1858.

Davies, P. *God and the New Physics.* New York: Simon & Schuster, 1983.

———. *The Cosmic Blueprint: New Discoveries in Nature's Creative Ability to Order the Universe.* New York: Simon & Schuster, 1988.

———. *The Mind of God: The Scientific Basis for a Rational World.* New York: Simon & Schuster, 1992.

———. *About Time: Einstein's Unfinished Revolution.* New York: Simon & Schuster, 1995.

Dawkins, R. *The Blind Watchmaker.* New York: W. W. Norton, 1986.

Dennett, D. *Consciousness Explained.* Boston: Little, Brown, 1991.

Descartes, R. *Discourse on Method and Meditations.* Trans. L. J. Lafleur. New York: Macmillan, 1960; originally published 1637 and 1641.

———. *Principles of Philosophy.* In *The Philosophical Works of Descartes,* ed. and trans. E. S. Haldane and G. R. T. Ross, 1:203–302. New York: Cambridge University Press, 1972; originally published 1644.

Dossey, L. *Space, Time and Medicine.* Boston: Shambhala, 1982.

———. "Where in the World Is the Mind?" *Advances* 6 (1989): 38–47.

———. "But Is It Energy? Reflections on Consciousness, Healing, and the New Paradigm." *Subtle Energies* 3, no. 3 (1992): 69–83.

———. *Healing Words: The Power of Prayer and the Practice of Medicine.* San Francisco: HarperCollins, 1993.

———. "Healing, Energy, and Consciousness: Into the Future or a Retreat to the Past?" *Subtle Energies* 5, no. 1 (1994): 1–33.

———. "Dialogue: But Is It Energy? Reflections on Consciousness, Healing, and the New Paradigm and Healing, Energy, and Consciousness: Into the Future or a Retreat to the Past?" *Subtle Energies* 5, no. 3 (1994): 264–71.

———. "Energy Talk." *Network: The Scientific and Medical Network Review* 63 (April 1997): 3–7.

———. "The Forces of Healing: Reflections on Energy, Consciousness, and the Beef Stroganoff Principle." *Alternative Therapies* 3, no. 5 (1997): 8–14.

Eckhart, M. *Meister Eckhart: Selected Writings.* Trans. O. Davies. New York: Penguin, 1994.

Einasto, J., M. Einasto, S. Gottlober, V. Muller, V. Saar, A. A. Starobinsky, E. Tago, D. Tucker, H. Andernach, and P. Frisch. "A 120-Mpc Periodicity in the Three-Dimensional Distribution of Galaxy Superclusters." *Nature* 385 (1997): 139–40.

Einstein, A. *Ideas and Opinions.* New York: Wings Books, 1954.

Einstein, A., and L. Infeld. *The Evolution of Physics.* New York: Simon & Schuster, 1938.

Emerson, R. W. *The Works of Ralph Waldo Emerson.* New York: Walter J. Black, n.d.

Faraday, M. *The Forces of Matter.* Buffalo, N.Y.: Prometheus Books, 1993.

Fawcett, J. "From a Plethora of Paradigms to Parsimony in Worldviews." *Nursing Science Quarterly* 6 (1993): 56–58.

———. "On the Requirements for a Metaparadigm: An Invitation to Dialogue." *Nursing Science Quarterly* 9 (1996): 94–97.

Feldman, T. S. "Energy." In *Dictionary of the History of Science,* ed. W. F. Bynum, E. J. Browne, and R. Potter, 122–23. Princeton, N.J.: Princeton University Press, 1981.

Feynman, R. P. *Six Easy Pieces.* Reading, Mass.: Addison-Wesley, 1995.

Field, S. "The Scientific Art of Medical Practice." In *Patterns of Rogerian Knowing,* ed. M. Madrid, 267–84. New York: National League for Nursing, 1997.

Flanagan, O. *The Science of the Mind.* 2d ed. Cambridge, Mass.: MIT Press, 1984.

Fox, M., ed. *Illuminations of Hildegard of Bingen.* Santa Fe, N.M.: Bear & Co., 1985.

Fraser, J. T., ed. *The Voices of Time.* New York: George Braziller, 1966.

Gadamer, H. *The Enigma of Health.* Stanford, Calif.: Stanford University Press, 1996.

Gallagher, K. T. " 'Natural Selection': A Tautology?" *International Philosophical Quarterly* 29 (1989): 17–31.

Gibbs, W. W. "A Massive Discovery." *Scientific American* (August 1998): 18, 20.

Gill, M. L. *Aristotle on Substance: The Paradox of Unity.* Princeton, N.J.: Princeton University Press, 1989.

Glanz, J. "Exploding Stars Point to a Universal Repulsive Force." *Science* 279 (1998): 651–52.

———. "Astronomers See a Cosmic Antigravity Force at Work." *Science* 279 (1998): 1298–99.

———. "Has a Dark Particle Come to Light?" *Science* 283 (1999): 13–14.

Gleick, J.. *Chaos: Making a New Science.* New York: Penguin, 1987.

Gold, J. L. "Practicing Medicine in the Nineties with an Emphasis on the Unitary Perspective of Patient Care." In *Patterns of Rogerian Knowing,* ed. M. Madrid, 257–66. New York: National League for Nursing, 1997.

Gribbin, J. *In Search of the Big Bang.* New York: Bantam, 1986.

Guthrie, W. K. C. *The Greek Philosophers: From Thales to Aristotle.* New York: Harper & Row, 1950.

Haisch, B., A. Rueda, and H. E. Puthoff. "Beyond $E=mc^2$." *The Sciences* 34 (1994): 26–31.

Harman, P. M. *Energy, Force, and Matter.* New York: Cambridge University Press, 1982.

Harvey, A., ed. *The Essential Mystics: The Soul's Journey into Truth.* New York: HarperCollins, 1996.

Hawking, S. W. *A Brief History of Time: From the Big Bang to Black Holes.* New York: Bantam, 1988.

Heisenberg, W. *Physics and Philosophy: The Revolution in Modern Science.* New York: Harper & Row, 1958.

———. *Physics and Beyond: Encounters and Conversations.* Trans. A. J. Pomerans. New York: Harper & Row, 1971.

Herbert, N. *Quantum Reality: Beyond the New Physics.* New York: Anchor Books, 1985.

Hesse, M. "Ether." In *Encyclopedia of Philosophy,* ed. P. Edwards, 2:66–69. New York; Macmillan and Free Press, 1967.

Hick, J. *The Existence of God.* New York: Macmillan, 1964.

Himmelfarb, G. *Darwin and the Darwinian Revolution.* New York: Anchor Books, 1962.

Hippocrates. "Ancient Medicine." In *Hippocrates,* trans. W. H. S. Jones, 1:13–64. Cambridge, Mass.: Harvard University Press, 1923.

———. "Airs, Waters, Places." In *Hippocrates,* trans. W. H. S. Jones, 1:66–137. Cambridge, Mass.: Harvard University Press, 1923.

———. *The Medical Works of Hippocrates.* Trans. J. Chadwick and W. N. Mann. Oxford: Blackwell Scientific Publications, 1950.

———. "Epidemics," Book 6. In *Hippocrates,* trans. W. Smith, 7:219–91. Cambridge, Mass.: Harvard University Press, 1994.

Hogan, C. J., R. P. Kirshner, and N. B. Sunzeff. "Surveying Space-Time with Supernovae."*Scientific American* 280, no. 1 (1999): 46–51.

Horgan, J. "Science Set Free from Truth." *New York Times,* July 16, 1996, A17.

Hume, D. *On Human Nature and the Understanding.* Ed. A. Flew. New York: Macmillan, 1962.

Inglis, B. *A History of Medicine.* New York: World, 1965.

Ivanhoe, P. J. "Ch'i." In *The Cambridge Dictionary of Philosophy,* ed. R. Audi, 115. New York: Cambridge University Press, 1995.

———. "li." In *The Cambridge Dictionary of Philosophy,* ed. R. Audi, 434. New York: Cambridge University Press, 1995.

Jackomos, A., and D. Fowell. *Aboriginal Living Histories: Stories in the Oral Tradition.* Cambridge: Cambridge University Press, 1992.

Jaffe, B. *Actualism.* November 1993. Available at http://www.actualism.org and via e-mail at brjaffe@actualism.org

James, W. *William James: Writings 1902–1910.* New York: Library of America, 1987.

Jammer, M. "Energy." In *Encyclopedia of Philosophy,* ed. P. Edwards, 1:511–17. New York: Macmillan and Free Press, 1967.

Jantsch, E. *The Self-Organizing Universe: Scientific and Human Implications of the Emerging Paradigm of Evolution.* Oxford: Pergamon Press, 1980.

Jastrow, R. *Red Giants and White Dwarfs: The Evolution of Stars, Planets and Life.* New York: Harper & Row, 1967.

Johnson, G. *Fire in the Mind: Science, Faith, and the Search for Order.* New York: Vintage, 1996; originally published 1995.

———. "Dark Matter Lights the Void." *New York Times Week in Review,* January 21, 1996, 1, 4.

———. "The Real Star Wars: Between Order and Chaos." *New York Times,* January 19, 1997, E4.

———. "Researchers on Complexity Ponder What Exactly They Study." *New York Times,* May 6, 1997, C1, C7.

———. "The Unspeakable Things That Particles Do." *New York Times,* July 27, 1997, E5.

———. "Once upon a Time, There Was a Big Bang Theory." *New York Times (On the Web),* www.nytimes.com, March 8, 1998, 1–4.

———. "Elusive Particles Continue to Puzzle Theorists of the Sun." *New York Times (On the Web),* www.nytimes.com, June 9, 1998, 1–6.

Joule, J. P. "On Matter, Living Force, and Heat." In J. P. Joule, *The Scientific Papers of James Prescott Joule,* 265–76. London: Taylor & Francis, 1884; originally published 1847.

Jung, C. *The Archetypes and the Collective Unconscious.* Princeton, N.J.: Princeton University Press, 1969.

———. *Synchronicity: An Acausal Connecting Principle.* Trans. R. F. C. Hull. Princeton, N.J.: Princeton University Press, 1973; originally published 1955.

———. *The Portable Jung.* Trans. R. F. C. Hull, ed. J. Campbell. New York: Viking, 1971; originally published 1951.

Kaptchuk, T. J. "Historical Context of the Concept of Vitalism in Complementary and Alternative Medicine." In *Fundamentals of Complementary and Alternative Medicine,* ed. M. S. Micozzi, 35–48. New York: Churchill Livingstone, 1996.

Kauffman, S. *At Home in the Universe: The Search for Laws of Self-Organization and Complexity.* Oxford: Oxford University Press, 1995.

Keen, W. W. *I Believe in God and in Evolution.* Philadelphia: J. B. Lippincott, 1922.

Koestler, A. *The Roots of Coincidence.* New York: Random House, 1972.

Krauss, L. M. "Cosmological Antigravity." *Scientific American* 280, no. 1 (1999): 52–59.

Krieger, D. *Accepting Your Power to Heal: The Personal Practice of Therapeutic Touch.* Santa Fe, N.M.: Bear & Co., 1993.

———. "Higher-Order Emergence of the Self during Therapeutic Touch." In *Spiritual Healing,* ed. D. Kunz, 262–71. New York: Theosophical Publishing House, 1995.

Kuhn, T. S. "Energy Conservation as Simultaneous Discovery." In *Critical Problems in the History of Science,* ed. M. Clagett, 321–56. Madison: University of Wisconsin Press, 1959.

Kunz, D., ed. *Spiritual Healing.* New York: Theosophical Publishing House, 1995.

Kunz, D., and E. Peper. "Fields and Their Clinical Implications." In *Spiritual Healing,* ed. D. Kunz, 213–61. New York: Theosophical Publishing House, 1995.

Lao Tzu. *Tao Te Ching.* Trans. D. C. Lau. New York: Penguin, 1963.

Latta, R., trans. *Leibniz: The Monadology and Other Philosophical Writings.* London: Oxford University Press, 1898.

Leonard, G. *The Silent Pulse: A Search for the Perfect Rhythm That Exists in Each of Us.* New York: E. P. Dutton, 1978.

Lear, J. *Aristotle: The Desire to Understand.* Cambridge: Cambridge University Press, 1988.

Leibniz, G. *The Monadology.* In *Leibniz: The Monadology and Other Philosophical Writings,* trans. R. Latta, 215–71. London: Oxford University Press, 1898; originally published 1714.

LeShan, L. *The Medium, the Mystic, and the Physicist: Toward a General Theory of the Paranormal.* New York: Penguin, 1966.

Leskowitz, E. "Dialogue: But Is It Energy? Reflections on Consciousness, Healing, and the New Paradigm and Healing, Energy, and Consciousness: Into the Future or a Retreat to the Past?" *Subtle Energies* 5, no. 3 (1994): 259–63.

Leslie, J., ed. *Physical Cosmology and Philosophy.* New York: Macmillan, 1990.

Levine, M. "The Four Conservation Principles of Nursing." *Nursing Forum* 6 (1967): 45–56.

———. "The Conservation Principles: A Retrospective." *Nursing Science Quarterly* 1 (1996): 38–41.

Lewen, R. "Punctuated Equilibrium Is Now Old Hat." *Science* 214 (1986): 672.

Linde, A. "The Self-Reproducing Universe." *Scientific American: Magnificent Cosmos* 9, no. 1 (1998): 98–104.

Lindsay, R. B., ed. *Energy: Historical Development of the Concept.* Stroudsburg, Pa.: Dowden, Hutchinson & Ross, 1975.

———. *Applications of Energy: Nineteenth Century.* Stroudsburg, Pa.: Dowden, Hutchinson & Ross, 1976.

Lodge, O. *Energy.* New York: Robert M. McBride, 1929.

Lorenz, E. *The Essence of Chaos.* Seattle: University of Washington Press, 1993.

Lovelock, J. E. *Gaia: A New Look at Life on Earth.* New York: Oxford University Press, 1979.

Lucretius. *Of the Nature of Things.* Trans. W. E. Leonard. New York: E. P. Dutton, 1921.

Lusthaus, D. "Laozi (Lao Tzu)." In *Great Thinkers of the Eastern World,* ed. I. P. McGreal, 9–15. New York: HarperCollins, 1995.

Macrae, J. *Therapeutic Touch: A Practical Guide.* New York: Alfred A. Knopf, 1987.

———. "Nightingale's Spiritual Philosophy." *Image: Journal of Nursing Scholarship* 27 (1995): 8–10.

———. "Therapeutic Touch as Meditation." In *Spiritual Healing,* ed. D. Kunz, 272–88. Wheaton, Ill.: Quest Books, 1995.

Madrid, M. "Participating in the Process of Dying." In *Rogers' Scientific Art of Nursing Practice,* ed. M. Madrid and E. A. M. Barrett, 91–103. New York: National League for Nursing, 1994.

Madrid, M., ed. *Patterns of Rogerian Knowing.* New York: National League for Nursing, 1997.

Madrid, M., and E. A. M. Barrett, eds. *Rogers' Scientific Art of Nursing Practice.* New York: National League for Nursing, 1994.

Magner, L. N. *A History of Medicine.* New York: Marcel Dekker, 1992.

Maimonides, M. *The Essential Maimonides: Translations of the Rambam.* Northvale, N.J.: Jason Aronson, 1996.

Malinski, V. M., ed. *Explorations on Martha Rogers' Science of Unitary Human Beings.* Norwalk, Conn.: Appleton-Century-Crofts, 1986.

Margenau, H. "Science, Creativity, and Psi." In *Psi Factors in Creativity,* ed. A. Angoff and B. Shapin, 78–84. New York: Parapsychology Foundation, 1970.

Maslow, A. H. *Toward a Psychology of Being.* New York: Van Nostrand Reinhold, 1968.

———. *The Farther Reaches of Human Nature.* New York: Viking, 1971.

Maxwell, J. C. *A Dynamical Theory of the Electromagnetic Field.* Edinburgh: Scottish Academic Press, 1982; originally published 1865.

May, J. G. "Lines of Force." In *Dictionary of the History of Science,* ed. W. F. Bynum, E. J. Browne, and R. Potter, 239–40. Princeton, N.J.: Princeton University Press, 1984.

Mayer, J. R. "On the Forces of Inorganic Nature." In *Men of Physics, Julius Robert Mayer: Prophet of Energy,* ed. R. B. Lindsay, 67–74. New York: Pergamon Press, 1973; originally published 1842.

McEvoy, M. D. "The Relationships among the Experiences of Dying, the Experience of Paranormal Events, and Creativity in Adults." In *Visions of Rogers' Science-Based Nursing,* ed. E. A. M. Barrett, 209–28. New York: National League for Nursing, 1990.

McGreal, I. P., ed. *Great Thinkers of the Eastern World.* New York: HarperCollins, 1995.

Meehan, T. C. "Therapeutic Touch and Postoperative Pain: A Rogerian Research Study." *Nursing Science Quarterly* 6 (1993): 69–78.

Micozzi, M. S., ed. *Fundamentals of Complementary and Alternative Medicine.* New York: Churchill Livingstone, 1996.

Mitchell, E. D. *The Way of the Explorer: An Apollo Astronaut's Journey through the Material and Mystical Worlds.* New York: G. P. Putnam's Sons, 1996.

Monroe, C., D. M. Meekhof, B. E. King, and D. J. Wineland. "A 'Schrodinger Cat' Superposition State of an Atom." *Science* 272 (1996): 1131–36.

Moody, R. A. *Life after Life: The Investigation of a Phenomenon—Survival of Bodily Death.* New York: Bantam, 1975.

Moran, G. *A Grammar of Responsibility.* New York: Crossroad, 1996.

Myss, C. *Anatomy of the Spirit: The Seven Stages of Power and Healing.* New York: Harmony Books, 1996.

Newman, M. "Prevailing Paradigms in Nursing." *Nursing Outlook* 40 (1992): 10–13, 32.

Newton, I. *Newton's Optical Writings: A Guided Study.* Trans. D. L. Sepper. New Brunswick, N.J.: Rutgers University Press, 1994.

———. *The Principia.* Trans. A. Motte. New York: Prometheus Books, 1995; originally published 1704.

Nightingale, F. *Notes on Nursing: What It Is, and What It Is Not.* Facsimile of 1st ed. (1859). London: Harrison and Sons, 1946.

———. *Suggestions for Thought.* Ed. M. D. Calabria and J. A. Macrae. Philadelphia: University of Pennsylvania Press, 1994; originally published 1860.

Normile, D. "Weighing in on Neutrino Mass." *Science* 279 (1998): 1698–90.

O'Manique, J. *Hyperphysics: Energy in Evolution.* New York: Humanities Press, 1969.

O'Regan, B. "Healing, Remission and Miracle Cures." In *Noetic Sciences Collection 1980–1990: Ten Years of Consciousness Research*, ed. B. McNeill and C. Guion, 44–54. Sausalito, Calif.: Institute of Noetic Sciences, 1991.

Overbye, D. "A Famous Einstein 'Fudge' Returns to Haunt Cosmology." *New York Times (On the Web)*, www.nytimes.com, May 26, 1998, 1–6.

Paley, W. "The Watch Analogy." In *Natural Theology*, ed. F. Ferre, 81–85. Indianapolis: Library of Liberal Arts, 1963.

Parmenides. *Parmenides: A Text with Translation, Commentary and Critical Essays*. Trans. L. Taran. Princeton, N.J.: Princeton University Press, 1965.

Parse, R. R. *Major Paradigms, Theories, and Critiques*. Philadelphia: Saunders, 1987.

✓ Peat, F. D. *Synchronicity: The Bridge between Matter and Mind*. New York: Bantam, 1987.

Pegis, A., ed. *Introduction to St. Thomas Aquinas*. New York: Modern Library, 1948.

Pelikan, J., ed. *The World Treasury of Religious Thought*. Boston: Little, Brown, 1990.

Pellegrino, C. R., and J. F. Stoff. *Darwin's Universe: Origins and Crises in the History of Life*. 2d ed. Blue Ridge Summit, Pa.: TAB Books, 1986.

Penrose, R. *The Large, the Small and the Human Mind*. New York: Cambridge University Press, 1997.

Phillips, J. R. "Changing Human Potentials and Future Visions of Nursing: A Human Field Image Perspective." In *Visions of Rogers' Science-Based Practice*, ed. E. A. M. Barrett, 13–25. New York: National League for Nursing, 1990.

———. "Search in Research." *Nursing Science Quarterly* 5 (1992): 50–51.

———. "The Open-Ended Nature of the Science of Unitary Human Beings." In *Rogers' Scientific Art of Nursing Practice*, ed. M. Madrid and E. A. M. Barrett, 11–25. New York: National League for Nursing, 1994.

Planck, M. *The Universe in the Light of Modern Physics*. New York: W. W. Norton, 1931.

Plato. *Dialogues of Plato*. Trans. B. Jowett. New York: Bigelow, Brown, n.d.

Polanyi, M. *Personal Knowledge: Towards a Post-Critical Philosophy*. Chicago: University of Chicago Press, 1962; originally published 1958.

Polkinghorne, J *Beyond Science: The Wider Human Context*. Cambridge University Press, 1996.

Poynting, J. H. "On the Transfer of Energy in the Electromagnetic Field." In *Applications of Energy: Nineteenth Century*, ed. R. B. Lindsay, 298–307. Stroudsburg, Pa.: Dowden, Hutchinson & Ross, 1976; originally published 1885.

Pribram, K. H. "Some Dimensions of Remembering: Steps Towards a Neuropsychological Model of Memory." In *Brain and Behavior*, vol. 2: *Perception and Action*, ed. K. H. Pribram, 252–58. Middlesex, England: Penguin, 1969; originally published 1966.

Prigogine, I. *From Being to Becoming: Time and Complexity in the Physical Sciences*. San Francisco: Freeman, 1980.

———. *The End of Certainty: Time, Chaos, and the New Laws of Nature*. New York: Free Press, 1996.

Prigogine, I., and I. Stengers. *Order Out of Chaos*. New York: Bantam, 1984.

Quinn, J. F. "Therapeutic Touch as Energy Exchange: Testing the Theory." *Advances in Nursing Science* 6 (1984): 42–49.

———. "Therapeutic Touch as Energy Exchange: Replication and Extension." *Nursing Science Quarterly* 2 (1989): 79–87.

Radhakrishnan, S., and C. A. Moore, eds. *A Sourcebook in Indian Philosophy.* Princeton, N.J.: Princeton University Press, 1957.

Redfield, J. *The Celestine Prophecy: An Adventure.* New York: Warner Books, 1993.

Reichenbach, H. *From Copernicus to Einstein.* Trans. R. B. Winn. New York: Dover, 1942.

Reinsel, R. "Parapsychology: An Empirical Science." In *Philosophy of Science and the Occult,* ed. P. Grim, 151–68. Albany: State University of New York Press, 1982.

Rhine, J. B. *Extrasensory Perception.* Boston: Bruce Humphries, 1934.

———. "Extrasensory Perception." In *Handbook of Parapsychology,* ed. B. B. Wolman, 163–74. New York: Van Nostrand Reinhold, 1977.

Ring, K. *Life at Death.* New York: Coward, McCann & Geoghegan, 1980.

———. "Near-Death Experiences: Implications for Human Evolution and Planetary Transformation." *Revision* 8 (1986): 75–85.

Rogers, M. E. *An Introduction to the Theoretical Basis of Nursing.* Philadelphia: F. A. Davis, 1970.

———. "Nursing: A Science of Unitary Man." In *Conceptual Models for Nursing Practice,* 2d ed., ed. J. P. Riehl and C. Roy, 329–31. New York: Appleton-Century-Crofts, 1980.

———. Nursing: "A Science of Unitary Human Beings." In *Conceptual Models for Nursing Practice,* 3d ed., ed. J. P. Riehl-Sisca, 181–88. Norwalk, Conn.: Appleton & Lange, 1989.

———. "Nursing Science and the Space Age." *Nursing Science Quarterly* 5 (1992): 27–34.

———. "The Science of Unitary Human Beings: Current Perspectives." *Nursing Science Quarterly* 7 (1994): 33–35.

Rudd, P. M. "Abstract: Science and Spirituality: A Personal View." Center for Theology and the Natural Sciences, Berkeley, Calif., Science and the Spiritual Quest Conference, June 1998.

Russell, E. W. "The Fields of Life." In *Future Science: Life Energies and the Physics of Paranormal Phenomena,* ed. J. White and S. Krippner, 59–72. New York: Anchor Books, 1977.

Russell, P. *The Awakening Earth: The Global Brain.* London: ARK Paperbacks, 1982.

Sagan, C. *Contact.* New York: Simon & Schuster, 1985.

———. *Broca's Brain.* New York: Random House, 1985.

Sarter, B. *The Stream of Becoming.* New York: National League for Nursing, 1988.

———. "Some Critical Philosophical Issues in the Science of Unitary Human Beings." *Nursing Science Quarterly* 2 (1989): 74–78.

Schlegel, R. "Time and Thermodynamics." In *The Voices of Time,* ed. J. T. Fraser, 500–526. New York: George Braziller, 1966.

Schrodinger, E. *What Is Life? With Mind and Matter.* New York: Cambridge University Press, 1967.

Seaward, B. L. "Alternative Medicine Complements Standard." *Health Progress* 75 (1994): 52–57.

Seelig, M. G. *Medicine: An Historical Outline.* Baltimore: Williams & Wilkins, 1931.

Sellmann, J. D. "Zhang Zai (Chang Tsai)." In *Great Thinkers of the Eastern World,* ed. I. P. McGreal, 108–10. New York: HarperCollins, 1995.

Selye, H. *The Stress of Life.* Rev. ed. New York: McGraw Hill, 1976.

Sepper, D. L. *Newton's Optical Writings: A Guided Study.* New Brunswick, N.J.: Rutgers University Press, 1994.

Sheldrake, R. *A New Science of Life: The Hypothesis of Formative Causation.* Los Angeles: J. P. Tarcher, 1981.

———. "Cause and Effect in Science: A Fresh Look." In *Noetic Sciences Collection 1980–1990,* ed. B. McNeill and C. Guion, 74–83. Sausalito, Calif.: Institute of Noetic Sciences, 1991.

Sigerist, H. E. *A History of Medicine.* Vol. 2: *Early Greek, Hindu, and Persian medicine.* New York: Oxford University Press, 1961.

Singh, S. "The Proof Is in the Neutrino." *New York Times (On the Web),* www.nytimes.com, June 16, 1998, 1–2.

Slater, V. "Toward an Understanding of Energetic Healing," part 1: "Energetic Structures." *Journal of Holistic Nursing* 13 (1995): 209–24.

———. "Toward an Understanding of Energetic Healing," part 2: "Energetic Processes." *Journal of Holistic Nursing* 13 (1995): 225–38.

Smart, N. "Hinduism." In *Encyclopedia of Philosophy,* ed. P. Edwards, 4:1–5. New York; Macmillan and Free Press, 1967.

Talbot, M. *The Holographic Universe.* New York: HarperCollins, 1991.

Taylor, A. E. *Aristotle.* New York: Dover, 1955.

Taylor, J. *Superminds.* New York: Warner Books, 1975.

Teilhard de Chardin, P. *The Phenomenon of Man.* Trans. B. Wall. New York: Harper & Row, 1959.

———. *The Divine Milieu.* New York: Harper & Row, 1960.

———. *Human Energy.* Trans. J. M. Cohen. London: William Collins Sons, 1969.

———. *The Heart of Matter.* New York: Harcourt Brace Jovanovich, 1978.

Thuan, T. X. *The Birth of the Universe: The Big Bang and After.* New York: Harry N. Abrams, 1992.

Tiller, W. "Dialogue: But Is It Energy? Reflections on Consciousness, Healing, and the New Paradigm and Healing, Energy, and Consciousness: Into the Future or a Retreat to the Past?" *Subtle Energies* 5, no. 3 (1994): 253–58.

Todaro-Franceschi, V. "Must Acausality Imply Unpredictability?" *Visions: The Journal of Rogerian Nursing Science* 4 (1995): 46–50.

———. "The Enigma of Energy: A Philosophical Inquiry." Doctoral dissertation, New York University, 1997. *Dissertation Abstracts International,* UMI #9819881.

Toolan, D., S.J. "Nature Is a Heraclitean Fire: Reflections on Cosmology in an Ecological Age." *Studies in the Spirituality of Jesuits* 23, no. 5 (1991): 1–44.

Turnbull, C. M. *The Forest People: A Study of the Pygmies of the Congo.* New York: Simon & Schuster, 1961.

Underhill, E. "Mysticism and Theology." In *The World Treasury of Religious Thought,* ed. J. Pelikan, 525–36. Boston: Little, Brown, 1990.

Van Ness, H. C. *Understanding Thermodynamics.* New York: Dover, 1969.

Watson, A. "Inflation Confronts an Open Universe." *Science* 279 (March 6, 1998): 1455.

Watson, L. *Supernature.* New York: Anchor Press/Doubleday, 1973.

Weber, R. *Dialogues with Scientists and Sages: The Search for Unity.* New York: Routledge & Kegan Paul, 1986.

Weil, A. *Spontaneous Healing.* New York: Alfred A. Knopf, 1995.

Weinberg, S. *The First Three Minutes.* London: Andre Deutsch, 1977.

West, M. *The World Is Made of Glass.* New York: William Morrow, 1983.

White, J., and S. Krippner, eds. *Future Science: Life Energies and the Physics of Paranormal Phenomena.* New York: Doubleday, 1977.

Whitehead, A. N. *Nature and Life.* London: Cambridge University Press, 1934.

———. "The New Reformation." In *The World Treasury of Modern Religious Thought,* ed. J. Pelikan, 206–17. Boston: Little, Brown, 1990.

Wilford, J. N. "Scientists Report Discovering 'Missing Mass' Component." *New York Times,* April 29, 1997, C3.

———. "In the Light of Dying Stars, Astronomers See Intimations of Cosmic Immortality." *New York Times (On the Web),* www.nytimes.com, April 21, 1998.

———. "Cosmologists Ponder 'Missing Energy' of the Universe." *New York Times (On the Web),* www.nytimes.com, May 5, 1998, 1–4.

Wilson, E. O. *The Diversity of Life.* Cambridge: Harvard University Press, 1992.

Wittgenstein, L. *Philosophical Investigations.* Trans. G. E. M. Anscombe. New York: Macmillan, 1953.

Wolman, B. B., ed. *Handbook of Parapsychology.* New York: Van Nostrand Reinhold, 1977.

Woodhouse, M. *Paradigm Wars: Worldviews for a New Age.* Berkeley, Calif.: Frog, 1996.

Wright, R. "Can Machines Think?" *Time,* March 25, 1996, 50–56.

Yoon, C. K. "Lake Victoria's Lightning-Fast Origin of Species." *New York Times,* August 27, 1996, C1, C4.

———. "Clues to Redwood's Mighty Growth Emerge." *New York Times (On the Web),* www.nytimes.com, November 24, 1998.

Ywahoo, D. *Voices of Our Ancestors: Cherokee Teachings from the Wisdom Fire.* Boston: Shambhala, 1987.

Zukav, G. *The Dancing Wu Li Masters: An Overview of the New Physics.* New York: Bantam New Age Books, 1979.

Zysk, K. "Traditional Ayurveda." In *Fundamentals of Complementary and Alternative Medicine,* ed. M. S. Micozzi, 233–42. New York: Churchill Livingstone, 1996.

Index